G000155891

The Irish Protestant Churches in the Twentieth Century

Alan Megahey

First published in Great Britain 2000 by
MACMILLAN PRESS LTD
Houndmills, Basingstoke, Hampshire RG21 6XS and London
Companies and representatives throughout the world

A catalogue record for this book is available from the British Library.

ISBN 0–333–73251–0

First published in the United States of America 2000 by
ST. MARTIN'S PRESS, LLC,
Scholarly and Reference Division,
175 Fifth Avenue, New York, N.Y. 10010

ISBN 0–312–23601–8

Library of Congress Cataloging-in-Publication Data
Megahey, Alan J.
The Irish Protestant churches in the twentieth century / Alan Megahey.
p. cm.
Includes bibliographical references and index.
ISBN 0–312–23601–8 (hardcover)
1. Protestant churches—Ireland—History—20th century. 2. Protestant churches—Northern Ireland—HIstory—20th century. 3. Ireland—Church history—20th century. 4. Northern Ireland—Church history—20th century. 5. Church of Ireland—History—20th century. 6. Presbyterian Church in Ireland—History—20th century. I. Title.

BX4839 .M44 2000
280'.4'094150904—dc21

00–033340

This book is printed on paper suitable for recycling and made from fully managed and sustained forest sources.

10 9 8 7 6 5 4 3 2 1
09 08 07 06 05 04 03 02 01 00

Printed and bound in Great Britain by
Antony Rowe Ltd, Chippenham, Wiltshire

IN MEMORY OF

my father

JACK MEGAHEY

Irish Protestant by birth

and

my father-in-law

FRED JEFFERY

Irish Protestant by adoption

Contents

Preface

The Irish Protestant Churches are well served by dedicated archivists and librarians, and I would especially thank Dr Raymond Refaussé and his staff at the Representative Church Body Library of the Church of Ireland in Dublin, Mr Alan McMillan at the Presbyterian Historical Society in Church House, Belfast, and the Revd Robin Roddie in the Wesley Historical Society Archive, Belfast. They welcomed me and helped me with unfailing courtesy. The staff of the Queen's University Library (Special Collections) were very helpful, as were the staffs in the Cambridge University Library, the British Library, and the Public Record Office in Belfast. My thanks are due also to Professor George Boyce, the Revd Sydney Callaghan, Mr Jack Gamble, Professor Keith Jeffery, and at Macmillan to Mr Tim Farmiloe and Ms Aruna Vasudevan. Needless to say, they bear no responsibility for anything in this book; nor does my wife, whose help has been invaluable – all those evenings spent sorting and filing!

My apologies to those who are offended when I sometimes use 'Ulster' to mean Northern Ireland, or Anglican to mean Church of Ireland, or Catholic to mean Roman Catholic (and vice versa). The varied usage is determined by aesthetic rather than political or theological considerations.

I would like to record my gratitude to the Revd Dr Owen Chadwick (sometime Professor of Modern History at Cambridge and formerly Master of Selwyn College), and to the Revd Dr Edward Norman (sometime Dean of Peterhouse, Cambridge, and now Chancellor of York Minster). While they were not even aware that this book was being written, they have been immensely helpful to me for more than thirty-five years. I cannot hope to match their fluency, profundity and scholarship, but I have been enormously enriched both by their friendship and by their books.

The Dedication is more than a filial duty; it is a tribute to two people whose benign influence and deep commitment is rather inadequately described by the term 'Irish Protestant'. They touched many lives.

A.J. Megahey

Introduction

Irish Protestants have been much written about – by historians, journalists and sociologists. Much attention, especially since the beginning of 'the Troubles' in Northern Ireland, has been given to 'Protestantism', and exhaustive historical and sociological – and some theological – discussion has analysed this 'settler community', these 'fearful people'. And indeed the wilder shores of Ulster Protestantism have attracted perhaps more than their fair share of attention. But the Protestant Churches themselves have been neglected. Perhaps 'church history' or 'ecclesiastical history' is regarded as *passé*. Perhaps it is too confusing, and too disruptive of analytical procedures, to examine institutions which, while central to the 'Northern Ireland problem', are all-Ireland in their dimensions.

This is a history of the Irish Protestant churches in the twentieth century. It is by no means exhaustive. All too many of the issues which are raised cannot be adequately treated within the compass of this book. Many themes deserve further exploration and fuller treatment: the role of Protestants overseas (as settlers and as missionaries); the history of Protestant secondary education in Ireland as a whole; the 'gospel hall' tradition; the influence of Protestantism on, or its symbiotic relationship with literature and art; the role of women in the churches. All these themes, and others, which are mentioned in this book, cry out for further examination.

Again, the perspective in which the Irish Protestant churches – or more usually 'Irish Protestants' – have been seen has often been skewed. Apart from the more obvious distortions – Marxist analysis for example – there has been a tendency to look at the Irish churches as if they ought to be like the English churches; that, somehow, modern English Christianity is normative. This is a viewpoint which, as recent Lambeth Conferences have shown, is not necessarily widely accepted even within world Anglicanism. A world view shaped by the Enlightenment, the industrial revolution, and the technological and social revolutions of the twentieth century has resulted in an analysis which has emphasised the growth of secularisation, and has increasingly, and often unconsciously, identified the Christian faith with liberalism, democracy and human rights. Peter Berger's *The Secular City*, published in 1966, well encapsulated the belief that modernisation will inevitably lead to secularisation. But as Cathal Daly pointed out in 1982, there has been much development and 'a change of perspective since the brash days of *The Secular City* and the "Death of God" theologising'.[1] Berger himself admitted in 1997 that he had been wrong to regard fundamentalism – or any 'passionate religious movement' – as a phenomenon increasingly rare in the modern world, because 'in fact it is not rare at all. What is rare is people who think otherwise. What is difficult to understand is not Iranian mullahs but American university professors.'[2] The influence of religious faith in the lives of the Irish, their use of religious terminology, and their high level of church-going, do not look so arcane when placed in that context, or in the context of major world developments in the late twentieth century. The role of Pope John Paul II in the fall of Communism, and of Christians like Archbishop Tutu in the fall of apartheid, seemed to suggest that religion – even institutional religion – was not necessarily a 'spent force'. In other words, the 'liberal humanist' interpretation of events in Ireland is not necessarily the most illuminating one, but it is the one which has been most in evidence, and is one which is likely to ignore, or at least to diminish, the role of organised religion as a positive force in the lives of ordinary people.

The place of religion in the lives of Protestant people in Ireland has sometimes been discussed in terms of comparison with other societies. Kurt Bowen has noted how 'Ulster's Protestant "popular religion" in many ways is like its American counterpart, amorphous, and at times badly divided, yet a power to be reckoned with.'[3] There were strong nineteenth-century American influences on English nonconformity and on Irish Protestantism – catchy songs, individual communion cups, teetotalism, 'American organs'.[4] There was an 'underlying common culture, a Protestant bond', a 'special relationship' which long pre-dated the one forged by Roosevelt and Churchill. It was a two-way process, as English and Irish nonconformists had, since the seventeenth century, emigrated to America, and helped to forge the United States. The American influence on Irish Protestantism continued throughout the twentieth century, especially among the smaller denominations and groups, many of whose halls or churches would not look out of place in the USA, whose pastors sometimes speak in what could be called a mid-Atlantic accent, and whose members might look enviously at areas of the USA where evolution may not be taught in schools or alcohol sold in shops. Sometimes the comparison is between Ulster Protestants and Afrikaners, with 'a confessional background in common',[5] both strongly influenced by Calvinism and both living within 'cultures that derive from the ancient Hebrew covenant'.[6] It looked very different at the beginning of the twentieth century. Then it was Irish republicans who 'identified with the Afrikaners' as fellow freedom fighters, while most Irish Protestants saw the Boers as a threat to the empire which they so fervently supported. Only latterly did republicans begin to sympathise with the 'previously ignored' blacks.[7] Unlike the links between Irish and American Protestantism, the comparison between the Ulster Protestants and the South African whites relies on the supposed similarity of the situation they each found themselves in: beleaguered, and bound together by history and faith. In the second half of the century the situation faced by the Rhodesian whites was also sometimes compared with that of the Ulster Protestants, and indeed right-wing Ulster opinion tended to support Ian Smith's illegal regime and thus reinforce a 'colonial' view of the Northern Ireland problem.[8] These comparisons tend to obscure as much as they reveal, especially given the extremely tiny white minority in Rhodesia, and the small white minority in South Africa, whereas Protestants are a majority of Northern Ireland's population, and a far larger minority in Ireland as a whole than the whites in South Africa. Ultimately, Irish Protestantism – which is, historically and structurally, more than merely 'Ulster' Protestantism – can only be understood by a closer examination of the Protestant churches themselves.

This book attempts to show how the Protestant churches operated as institutions, and also to show how the lives of their members were affected by, and in turn affected, those institutions. Sometimes the story must, of necessity, dwell on political events and government activity. It would be difficult to write about any aspect of twentieth-century Ireland with the politics taken out. Some great national events are intimately part of the story: the Ulster Covenant, two world wars, the creation of two jurisdictions in Ireland, and the outbreak of civil disorder in Northern Ireland are obvious milestones in the history of the churches, and in the lives of church members. But an attempt is made to keep the focus on the churches themselves, and on the members of the various churches.

Inevitably, the nearer one approaches the end of the century, the more difficult is the analysis of events from an historian's point of view. There are too many loose ends. There are too many outcomes not yet visible. Christianity has had a place in Irish history for over fifteen hundred years. Reformed Christianity has co-existed there with Roman Catholicism for almost a third of that time. At the end of the twentieth century a Presbyterian minister could ask whether 'we suspect that here we have no abiding city and that the Plantation will yet be reversed and we will, however reluctantly, return to the place whence we came 350 years ago, and Ireland shall no us no more?'[9] That 'precarious belonging', as he calls it, is a part of the story; as is the uncomfortable awareness of a decline in Protestant numbers in the south throughout the century. Cardinal Daly noted 'the suspicion of a Catholic church policy aimed at the attrition of Protestants' numbers, or at least welcoming the prospect of this attrition'. But he went on: 'The "five per cent" is not doomed to continued shrinkage. I thank God for that.'[10] No Catholic bishop would have uttered such sentiments at the beginning of the century. The response of many Protestants at the end of the century to those words would be one of suspicion: Catholic triumphalism dressed up in ecumenical vestments. Thus they, curiously, subscribe to a colonial view of the situation: as if the 'Protestant moment' in Irish history were – like the 'colonial moment' in the history of Africa – doomed to foreclosure. That analysis slips over into a political and constitutional one. Perhaps it was not so much Protestantism that was at risk; there was a feeling that the 'British way of life' was under threat, partly from militant republicanism, and partly from a British government looking for a way out of an age-old problem.[11] But the Protestants in Northern Ireland remained in a majority. And the churches in the Republic of Ireland – by the end of a century that had seen great attrition of numbers – did not seem to be in terminal decline. Numbers might even be edging upwards, just as the population of the island was increasing after a century and a half of decline.

The historian cannot predict. Nor can he tell the whole story. He can record the 'changes and chances of this fleeting world'; those indeed are the stuff of history. He can and must seek to explain them. The workings of Providence meanwhile, even in a work on the churches, are beyond his terms of reference.

1
The Churches in 1900 – An Anatomy

> Religion – it means everything over here – everything mostly but what it ought to mean. A kind of badge or uniform, that's what it is mostly, a sign of our ascendancy by right and force of law, something that marks and values us like a brand on a butter-tub. It puts us in classes, and keeps us there; just as if it was a curse laid upon us instead of God's blessing.
>
> *Shan Bullock's father, born 1840*[1]

A Protestant people

The census of 1901 revealed that there were 4.5 million people in Ireland. Because the Irish census (unlike that in the rest of the United Kingdom) contained a religious question, we know that three-quarters of the population were Roman Catholics. Virtually all the rest – 1,150,114 people – were Protestants. Proportionately, there were more Protestants in Ireland in 1901 than there had been 40 years earlier: 26% as against 22%. Numerically there were fewer, as the population of Ireland had fallen by over a million during that period; indeed it had been in decline since the famine in the 1840s. In 1831, before that dreadful event, Ireland's population had been over half that of England and Wales; now it was just over a tenth. And the Protestant population of the whole of Ireland was about the same as the total population of Staffordshire or County Durham.

Almost all of those Protestants were proud of their birthright as citizens of the United Kingdom, with its Protestant monarch, and its sense of nationality still largely determined by its Protestantism. As Linda Colley has vividly demonstrated, 'Protestantism lay at the core of British national identity.'[2] In 1895 an English bishop wrote of 'the strength and depth of the present Protestantism of England', and in words to delight Irish Protestants he declared 'that it is both strong and deep'.[3] That shared sense of a national identity defined in religious terms was important to Irish Protestants, who would retain their loyalty to it long after it had faded in England. They could also feel kinship with the great 'Anglo-Saxon nations' of the world. These years, at the end of the nineteenth century, 'witnessed the high point of American Protestant civilization'.[4] and there were strong links, Protestant as well as Catholic, between Ireland and America. The other significant 'Protestant' nation was Germany which, despite its large Catholic population, was usually seen as a Protestant country, with a Protestant monarch. Germany was, after all, the place that had given the world Martin Luther and the Reformation, and the British their royal family. Even during the Boer War Irish Protestants could note with approval that the Kaiser was a ruler who preached – and preached good Protestant sermons.[5]

So although there were only a million Irish Protestants, they could feel themselves part of something greater – a world-wide Protestant movement. And of course they were part, too, of British Protestantism. It was not merely a generalised affinity. All the Irish churches had strong links, informal or institutional, with their sister or parent churches in Britain. They could feel sustained by fellow-Protestants there. The English Wesleyan, the Revd C.H. Kelly, addressed the Free Church meeting in Sheffield in March 1900. 'With all the experience of long ages', he declared, 'Popery has not improved; with all the religious and civil liberty granted to its Church in this Empire, it is still the foe of true liberty, of free speech, of private judgment.'[6] Such sentiments were often heard on Irish Protestant platforms and from Irish pulpits; to hear them in England – as one still could in the early years of the twentieth century – ensured that Irish Protestants need not feel isolated.

But Ireland was also different. It was the only component part of the United Kingdom where there was no established church, and while the (Anglican) Church of Wales was still in line for disestablishment if the Liberals had their way, the religious establishments in England and Scotland seemed fairly secure. Ireland was the only part of the United Kingdom where Roman Catholics were in an overwhelming majority. Indeed it was virtually the only part of the British Empire (apart from

Malta) where this was the case. And that was not a mere denominational statistic but a crucial fact in Irish religious, political, intellectual and cultural life. For Ireland was the only part of the United Kingdom where, for almost everyone, religious affiliation, political allegiance and a sense of cultural and national identity all went hand in hand. And it was the only part of the United Kingdom where, as A.T.Q. Stewart has succinctly put it, violence 'would appear to be endemic'.[7] That 'most distressful country' had, throughout the nineteenth century, experienced armed uprisings, rural disorder, and urban conflict on a scale that set it apart from the rest of the United Kingdom. Other differences are evident too: Britain was a wealthy, prospering, urbanised society – despite the poverty and social inequalities of General Booth's 'Darkest England'.[8] Ireland retained (certainly outside Dublin and the extreme north east) most of the characteristics of a rural, backward, pre-industrial society.

So while Irish Protestants could feel part of British and world Protestantism, they also had to exist in a land very different from Britain, or America, or Germany, and their existence was sustained by the ministrations of their churches. While most of them were Protestants first, with their denominational allegiance coming second, the institutional churches were for many of them the focus of their cultural, spiritual and social lives. Even for those who never darkened a church's doors (and evidence suggests there were far fewer of those in Ireland than in the rest of the United Kingdom), the fact of their Protestantism was a significant part of their inheritance, and a determining factor in their attitudes. Youth organisations – a growth industry in late Victorian and Edwardian Britain and Ireland – had church affiliations. The temperance movement – a vast, populist plethora of organisations throughout the United Kingdom – received most of its inspiration, encouragement and following from members of the churches. Social activities, the Orange Order, the rhythm of life for many (perhaps most) Irish Protestants, all related to the institutional churches. And the church with the largest number of adherents was the Church of Ireland.

The Church of Ireland

Just over half of all Irish Protestants were Anglicans, members of a church which until 1871 had been part of the United Church of England and Ireland. In that year, it was disestablished and disendowed – an understandable result of the Gladstonian Liberal crusade to settle the Irish problem, and to promote greater civil and religious liberty. On 1 January 1871, as the Church of Ireland ceased to be the established church, the Anglicans of Londonderry gathered in their cathedral. They sang a hymn penned by their bishop's wife, Mrs C.F. Alexander, author of 'All things bright and beautiful' and 'There is a green hill far away' – two highly popular hymns for Victorians and for later generations. This less-enduring hymn, however, expressed their current frustrations and their fears:

> Look down, Lord of Heaven, on our desolation!
> Fallen, fallen, fallen is now our Country's crown,
> Dimly dawns the New Year on a churchless nation,
> Ammon and Amalek tread our borders down.

It was a new year in which Irish bishops no longer received summonses to attend the House of Lords, and were no longer appointees of the crown; they would in future be chosen by an electoral college. The finances of the church, and the compensation paid to it, were vested in the Representative Church Body, while in the governance of the church the archbishops and bishops shared authority not with the crown, but with an annual Synod, composed of clergy and laity, and in a sense predating by a century similar arrangements for the (still established) Church of England.

The archbishoprics in Ireland were the sees of Armagh (the primatial see) and Dublin. Two other archbishoprics (of Tuam and Cashel) had been abolished in 1833 by the Irish Church Temporalities Act, which had also reduced the number of Irish bishoprics from 20 to 12. Arguably, the effect of the act was greater in England than in Ireland, for it stimulated John Keble's assize sermon, which – as Newman always claimed – marked the start of the Oxford Movement, and the transformation of Anglicanism in terms of both doctrine and liturgy. Despite the Irish stimulus in this transformation, the Church of Ireland itself was not so transformed. Indeed it remained frozen in the Protestant Anglicanism of the 1860s. Its General Convention, which established the codes of conduct for the newly 'independent' church, laid down strict guidelines, ensuring the Protestant nature of the Church of Ireland, at a time when the English church was troubled by a rising tide of ritualism. The Canons of 1871, which were incorporated into the new constitution, erected strong defences against such ritualism. Eucharistic vestments were forbidden, as were the use of incense, the mixing of water with wine at communion, processions in the church, or a cross and candles on the communion table (not 'altar'). All these were issues that were to plague the Church of England for more than a quarter of a century thereafter. A new Prayer Book was issued finally in 1879, little different from the 1662 Book of Common Prayer save for the omission of the words 'I absolve thee' in the service for Visitation of the Sick (because of its 'priestly' connotations), and a clause no longer requiring the Athanasian Creed to be said in church (because of widespread reservations about the 'damnatory clauses' stating that persons could not be saved who did not adhere unwaveringly to the doctrines of the creed).[9] The result was that, as was claimed in 1886, 'high, low and broad are terms practically unknown to us'.[10] Or, as a modern commentator has put it, the Irish Church 'emerged from the revision controversy stronger than before: with its Catholicity intact; with firm (perhaps excessive) safeguards against ritualism embodied in its canons; and with a body of law which was understood, respected and obeyed. This last was more than could be said about the Church of England at that time or since.'[11]

Although it had been disestablished for 30 years at the start of the new century, the Church of Ireland still had some claims to being a 'national' church. Its parochial system – like the Roman Catholic parishes – covered the whole of Ireland, which was a comfort even to non-Anglican Protestants.[12] The reason for this network is of course historic – all the pre-Reformation churches and cathedrals of Ireland were retained by the Anglicans after disestablishment, much to the disappointment of Roman Catholics who had hoped for the 'return' of at least some of them. The sustaining of such a network was costly in terms of upkeep costs and manpower. There were more than two dozen cathedrals to maintain and man, two in Dublin alone. Thus the tiny Church of Ireland was burdened with almost the same number of cathedrals as the Church of England, whose buildings were admittedly much larger and grander. The Church of Ireland was 'top heavy' both in plant and in manpower. There was one Irish Anglican priest to every 359 Anglicans; by contrast, there was one Presbyterian minister to every 647 Presbyterians – and that even though Presbyterians were much more concentrated in the north east of the island.

Another aspect of its 'national church' heritage was its name. The Church of Ireland continued to carry that title, even after disestablishment, and despite attempts – particularly by Presbyterians – to call it the 'Episcopal Church'. For Irish Anglicans, it was more than just a name. 'We claim', said J.H. Bernard (later Archbishop of Dublin) in 1904, 'to be the representatives of the ancient church of our country; we have good historical proofs in support of our claim; our claim to the title "Church of Ireland" has been recognised by the law; we have maintained through good and ill report, our witness for the catholic Faith, against Protestant dissent on the one hand and against Roman novelties of belief and practice on the other.'[13] That is a succinct summary of the 'high church' aspect of Irish Anglicanism, with its emphasis on the unbroken historic episcopate. Its claim to the title 'Church of Ireland' continued for some time to have 'the same effect on a certain class of Presbyterian minister that a red rag has on a bull'.[14] But despite the scoffing – St John Ervine called the title 'the comical description given to themselves by the Episcopalians'[15] – the Church of Ireland continued, and continues to call itself and to be called by a name which still seems to suggest a national church.[16]

In 1914 there were 1,483 Anglican clergy in Ireland – some of them retired, though there was no retirement age at that time. The overwhelming majority (88%) were graduates, mostly of Trinity College, Dublin (TCD), and virtually all received their theological training there. Only 65 had attended Oxford or Cambridge, and – perhaps more surprisingly – only 86 were graduates of the Queen's Colleges (of Belfast, Cork or Galway). The comparative uniformity of the Church of Ireland, in terms of style and liturgy, is to be explained not only by its strict Canons, but also by the uniformity of the education received by its clergy. How different from the Church of England clergy who came from more variegated university and theological college backgrounds.

The education of the Anglican clergy at secondary school level demonstrates a similar uniformity. They attended what Bishop C.F. D'Arcy – exaggerating somewhat – called schools for the 'gentry', 'the royal schools of Armagh, Portora and Dungannon in the north; in the south Kilkenny College and Middleton College'.[17] These, along with Dublin High School, St Columba's College and Rathmines School educated a large proportion of the Anglican clergy. Rathmines School had a particularly notable record; in the years 1863 to 1900 its pupils included eleven future bishops: three of Tuam (Harden, Holmes and Ross), two of Meath (Keane and Collins) and one each of Kilmore (Moore), Down (Grierson), Killaloe (Patton), Cork (Dowse), Auckland, New Zealand (Neligan) and Chekiang, China (Curtis).[18] Later, in the twentieth century, the schools of the north became more important: Portora Royal School (Enniskillen) gave the church two bishops of Connor (Poyntz and Moore), one of Tuam (Crozier), one Archbishop of Dublin (Empey), and a school chaplain who preceded Empey at Dublin.[19] By the end of the twentieth century, the main 'feeder schools' for clergy were in Belfast, the Royal Belfast Academical Institution and the Methodist College between them leading the field.

In 1900, about a fifth of the Anglican clergy in Ireland had been ordained into the established church. Some could boast extraordinarily long careers. The Revd Canon J.L. Stawell resigned from the parish of Aghnameadle in Killaloe in 1911, having been rector there for 58 years. Ordained in 1842, his incumbency had begun before the famine. In 1900 in the diocese of Meath, 10 of the 80 active clergy had been ordained before disestablishment, the oldest being the Revd F.W. Wetheril of Rathmolyon, who had served the diocese for 47 years in the same parish. His income, in 1900, was £241 per annum. That was not untypical. The average salary for an incumbent was £265 per annum, although that average hides gross discrepancies. The Revd James Hannay, better known as the writer George A. Birmingham, describes an imaginary cleric in a 'miserable little West of Ireland living' earning £180, and indeed about a fifth of the livings in Ireland were worth less than £200.[20] By contrast J.H. Bernard became in 1888 the Archbishop King Lecturer at TCD with a salary of £900 per annum. The Bishop of Meath in the same year earned £1,500. Even the lowest incumbent's income however looks reasonable when compared with that of a secondary school master at £100 per annum, which was the same as a congregational minister's stipend.

Even in 1900, 30 years after disestablishment, there is a touch of Barsetshire about the Anglican clergy in Ireland. They were often men who moved easily between the rectory, the big house, the Dublin Club and the Trinity Combination Room. Looking back in 1906, an old Presbyterian minister remembered that when he started work in the west of Ireland the local incumbent 'was one of the high and dry school, an aristocrat who only visited among the gentry class, and that not as a pastor but as a friend'.[21] When in 1890 the Rector of Arklow got involved in a disturbance over street preaching, J.O. Hannay was dismayed; 'the local episcopal clergy were most of us gentlemen by birth and education', who had more dignity than to allow themselves to be 'hooted and stoned by the corner boys of a small provincial town'.[22] His comment exudes an 'ascendancy' feel, even though Hannay was brought up in a clerical household in industrial Belfast.

Although at the time of disestablishment over half of all Anglicans lived in Ulster, only a third of the representatives in the General Synod came from there. That disproportionate southern representation continued. By the 1960s, three-quarters of all Anglicans were in Ulster, but representation from the northern counties was still just over half. The Archbishop of Armagh, as was noted in 1961, he was better known in Dublin than to Ulstermen, who never experienced 'close proximity to the General Synod and all the ecclesiastical hub-bub that makes the

Primate of All Ireland a dominating figure in the life of the Church in Dublin'.[23] Alone among the Protestant churches the Anglicans retained Dublin as their administrative capital; it was not until 1986 that the General Synod met for the first time outside Dublin – in Belfast. It met there again in 1991, and in Cork in 1994. The church's educational centre continued to be Dublin, based on Trinity and its Divinity Hostel, and latterly also on the Theological College at Rathgar; and from 1884 the Church of Ireland (Teacher) training college was in Kildare Place in Dublin. The *Church of Ireland Gazette* was published in Dublin, its new name adopted in 1900 to replace the stuffier *Irish Ecclesiastical Gazette*. There was also a Belfast-based weekly called *The Warden* (from 1910 *The Irish Churchman*) which had a more Protestant and populist tone. In a sense it is more representative of the views of the majority of Irish Anglicans. In 1910, it carried an article entitled 'Why I am a Churchman'.[24] Fourteen reasons were given, among them these: 'Because I know of no Church that holds the great leading truths of the Gospel more simply'; 'Because our Church so signally honours the Bible'; 'Because I love the Protestant character of our Church.' These were all reasons for any Irish Protestant to declare allegiance to his particular church. And the emphasis on the Gospel and the Bible makes clear that the Church of Ireland shared with all other denominations an evangelical faith – what we might call a 'pan-Protestant' approach. Even the most distinctively 'Anglican' of the reasons given was couched in moderate and almost pan-Protestant language: 'Because ... the Primitive Church was regulated, and her ministers were ordained by Bishops and Presbyters appointed (there is reason to believe) by the Apostles themselves, so our own Church is likewise regulated ... .' Leave out the word 'Bishops' and this comment could have been made by a Presbyterian.

The Presbyterian Church

While just over half of all Irish Protestants were Anglicans (581,089), just under half (443,276) were Presbyterians. They were almost exclusively concentrated in the north east of the country. Of the 36 presbyteries into which congregations were grouped in 1900, eleven were outside the six counties of the north east, but contained only 14% of the Presbyterian population. Despite that, the *Witness* – the Belfast-based semi-official newspaper of the church – declared in June 1900: 'We are not to be styled the Presbyterian Church of Ulster.'[25] Nevertheless that was where most Presbyterians lived, and their church headquarters was as firmly based in Belfast as the Anglican was in Dublin. A splendid Church House in Scottish baronial style was opened in central Belfast in 1905, just a block or two away from the City Hall. It contained a great hall seating over 2,000 and suites of offices and other halls, the location for committees and boards and annual meetings of the General Assembly. The Presbyterian Theological College ('Assembly's') sat just across the road from the Queen's College, Belfast, as a focus for education provided by the Presbyterian Theological Faculty of Ireland, which had been incorporated in 1881. Another Presbyterian College – Magee, in Derry – founded in 1865, was not as a theological college as some had hoped, but a university college with a theological faculty. From 1906 it had a link with TCD, thus allowing at least some Presbyterian ordinands a chance to spend a brief spell in what was to many of them *terra incognita*. The Presbyterian Theological Faculty, incorporating staff from Assembly's and Magee, could award theology degrees both on examination results and *honoris causa*. Not until 1929 did Queen's University itself have a theological faculty, which celebrated its launch in ecumenical style by awarding honorary degrees to the Anglican and Roman Catholic primates, and to the Principal of Assembly's.[26]

The Presbyterian clergy were mostly graduates (87% as compared with 88% for the Anglicans). Generally they had been to Queen's College (later University), Belfast, and to Assembly's. Some spent a spell at one of the colleges or universities in Scotland. Their schooling is more difficult to ascertain. Certainly Campbell College, founded in Belfast in 1894 by prominent Presbyterians as a 'public school', never became the Presbyterian equivalent of the Anglican Rathmines or Portora. Of the 215 entrants in its first year, five became clergymen – three Presbyterian and two Anglican. Of the 230 entrants in the period 1910–14, again five were eventually ordained – three in the Anglican Church, one in the Methodist and one in the Presbyterian. A breakdown of the backgrounds of Presbyterian ordinands is available. It reveals that of the first 100 Assembly's students – in the middle of the nineteenth century – 76% came from the land; this proportion dropped to 52% in the later part of the nineteenth century, and in the first half of the twentieth century dropped again, to some 35%.[27] But a sense of land and place remained significant for a people who traced their Irish presence back to the settlements of the seventeenth century. The great nineteenth-century Presbyterian historian Thomas Hamilton wrote feelingly of the boats full of settlers crossing the sea from Scotland: 'Those old-fashioned boats carried the fortunes of Ulster.'[28] There the Presbyterians settled, and their close relationship with the land began. A century later the autobiography of another leading Presbyterian historian, J.M. Barkley, seemed rooted in the 'rich, congenial soil' of Presbyterianism and the Ulster landscape.[29] The sense of land and place is even more evident in the writings of a late-twentieth-century Moderator, John Dunlop, whose sense of 'precarious belonging' we have already noted.[30]

Until disestablishment, the Presbyterians had of course been non-conformists or dissenters. Thereafter, even though their church was on the same legal footing as the Anglican, they still harboured radical and 'dissenting' views about the 'many jolterheads among the Tory landlords of Ulster', as the *Witness* described them in 1881, noting that a 'deep, widespread and growing discontent with the system of Irish land tenure pervades the Presbyterian farmers of Ireland'.[31] Here was that sense of 'precarious belonging' – whether in terms of their hold on the land, or of their place in the social and political structure. While the home rule controversy of subsequent decades ensured that most Presbyterians abandoned their support for Gladstonian Liberalism and became Unionists, elements of Presbyterian radicalism lived on. It was a Moderator in the early twentieth century who could still claim that the 'web of hate' in Ireland had been 'woven out of three strands – rack-rents, tithes (weighted with religious intolerance) and minority rule'.[32] All these had been a source of grievance among Presbyterians as among Roman Catholics, and the blame was being firmly laid at the door of the Anglican and aristocratic ascendancy. But the Presbyterian sense of history did not merely stretch back to the years before disestablishment, or two centuries earlier to the plantation of Ulster. Just as Anglicans had their doctrine of apostolic succession, so too the Presbyterians looked back to the primitive church of the first century. They believed that 'the original deposit of saving truth has been handed down to us, and is preserved in our Church', that 'the original Divine constitution of Christ's Church was Presbyterian, and so continued for three hundred years', this primitive constitution having been 'restored at the reformation, and adopted by all the best "reformed Churches" Continental and British'.[33] Here was a Presbyterian interpretation of church history that had no place for bishops and priests, but rather relied upon 'the right of the individual to approach God without a man-priest, and the right of the congregation to elect its own office-bearers'. Their church was governed not by clergy, but by the General Assembly. 'Presbyterians take their Reverend Court very seriously ... The General Assembly under God is the authority and its will and findings must be taken seriously.'[34] This Assembly, composed of clergy and laymen and meeting for a week each June, wielded enormous power. While in the Church of Ireland there was little interference in the affairs of local congregations (unless the canons of the church were being infringed), in the Presbyterian Church the local congregations were subject to official visitations every seven years, and any irregularities were reported to the Assembly. Presiding over the General Assembly was the Moderator, elected annually. As a late-twentieth-century Moderator pointed out, this democratic order has its disadvantages: 'People in Ireland know the names of the Catholic Cardinal and the Anglican Archbishop; but few ... could immediately name any current Moderator.'[35] Nevertheless for Presbyterians he is an important figure, 'sustained by a loyalty and treated with a deference which an Archbishop might envy'.[36] The respect due to him in the wider world became an issue at the beginning of the twentieth century. Precedence at state functions at the Vice-regal Lodge or at Dublin Castle was denied him at an appropriate level in 1902, leading to a fierce newspaper attack on 'The Castle', and on the 'fawning and cringing' to government which had brought no result. A Conservative government was in power, so a little of the old Presbyterian radicalism is evident here.[37] Within a year however the issue was resolved, with the Moderator given precedence with the Anglican Archbishops instead of the previous 'galling feature that the head of a great Church like ours had to accept a position behind that of a mere diocesan bishop'.[38] Thus, 30 years after disestablishment, and in a new century, the implications of the dismantling of the old confessional state were still being worked out.

Presbyterian clergy were on average less well paid than their Anglican counterparts. Around the turn of the century, the average salary in a rural presbytery was about £156 per annum, though in a Belfast church some £350.[39] It was even reported that the minister of the prestigious May Street Church in the city received 'considerably over £800 at least'.[40] These salaries were funded out of the local church's income, though with some help from central funds. At disestablishment, the Presbyterians lost the 'Regium Donum', a state subsidy for the Presbyterian clergy initially instituted under Charles II. This curious anomaly – a state subsidy for a church that was not the established church – amounted on the eve of disestablishment to just over £69 per minister.[41] The state payoff in 1869 was some £586,000, most of which was paid into a Sustentation Fund which still subsidised the salary of each clergyman.[42] Salary, however, is not much related to the 'call' to the ministry; up-bringing has much to do with it. Of the 2,114 students attending Assembly's in the years 1846–1953, 18% were sons of the manse.[43] Often they followed their fathers to occupy the very manse in which they had been born. J.A. Rentoul is a good example of this. He died in London in 1919, having been a QC and a Unionist MP. But he had started his career following his father as minister of 2nd Ray, in Donegal, where his grandfather had been ordained in 1791, and where his maternal great-grandfather, the Revd Robert Reid, had been the first minister in 1752. Rentoul's two uncles were also clergymen (of a tenant right persuasion, like his father). One of them, James (minister of Garvagh) had four sons who were Presbyterian clergymen, and one of those sons (Robert, minister of Clonmel) had two sons who were ministers. Rentoul's other uncle, John (minister of Ballymoney, and predecessor of J.B. Armour, the leading Presbyterian home ruler), had two sons who were ministers. It was an astonishing dynasty. A family less extensive but which in the twentieth century made a significant impact on the church was that of the Revd William Alexander Park, minister at Glendermott from 1911 until 1930. He and his wife (herself a daughter of a manse) had three sons: Samuel James, minister of Dun Laoghaire for 29 years and Moderator in 1965; William Alexander Albert, minister of Ballygilbert for 37 years and Moderator in 1961; John Ferguson, minister of Stormont for 39 years. All three men were awarded DDs by the Presbyterian Theological Faculty.

The smaller churches

The two great churches – Presbyterian and Anglican – dominate the history of Protestantism in Ireland in the twentieth century. The most significant of the smaller churches was the Methodist, which in 1901 had 61,979 members, some 5% of the Protestant population. The foundations of Methodism in Ireland had been laid by John Wesley himself, who paid no fewer than 21 visits to the country. As in England, Methodism in Ireland began as a ginger group within the established church. As it developed into an autonomous church, it suffered splits in its membership. The two largest Irish Methodist denominations, the Wesleyans and the Primitives, united in 1878 into one Methodist Church, over half a century before Methodist union in England. But the relationship between Methodists and Anglicans continued to be reflected in their geographical distribution. A study of the 1881 census figures, for instance, reveals that where the Church of Ireland was strong – as in Dublin, Wicklow, Antrim, Armagh, Down, Fermanagh and Tyrone, there was a strong Methodist element, and where it was weak – as in Clare, Galway and Mayo, the Methodists too were weak.[44] Although emigration, and the growth of Belfast, blurred this distribution pattern towards the end of the century, traces remained, especially in the south and west where Methodists were often the most biggest denomination after the Anglicans. It remained in another way too. In the 1950s, a Methodist minister in Co. Fermanagh discovered that a young man in his preparation class was also attending confirmation classes at the Anglican church, it being a 'tradition in that family to hold membership in both churches'.[45]

The Methodist Church retained much closer links with its sister (or parent) church in England than the Anglicans did with theirs, or the Presbyterians with those in Scotland. The President of the English Wesleyan (later, Methodist) Church presided at the annual Irish Conference, and continues to do so. From 1883 an Irishman held the post of vice-president, and was in effect the President of the Methodist Church (as opposed to Conference) in Ireland, though that title did not come into use until 1921. J.H. Rigg, the Victorian Methodist divine, once noted how the Wesleys 'recognised the remarkable reproduction in their own society of primitive and apostolic fellowship'.[46] In 1890 the Methodist semi-official newspaper, the *Christian Advocate*, noted that one of 'the features of Methodism which strikes outsiders most forcibly is what we sometimes speak of as "a godly sociality" ... one of the distinctive marks of Methodism'.[47] Such fellowship was understandable in the early days of the Methodist movement; that it continued was in part due to a number of distinctive features of the church. One was the itinerancy of ministers. The stationing committee of the conference re-deployed clergy every three years, and although by the late twentieth century their average stay in a church was more like five years, the obituaries of Methodist ministers still speak of their having 'travelled' for so many years – that is having been in the ministry for that number of years. This system, whatever its disruptive aspects, undoubtedly had the effect of knitting the 'connexion', as the Methodists called their church, more closely together. Another distinctive institution which did the same at a local level was the class meeting, whereby each congregation (or 'society' as it was tellingly called) was broken up into small groups for regular meetings in the church or in a private home, for prayer, study and worship. The Conference of 1905 declared that 'non-attendance at the meeting of the class does not of itself disqualify for membership of the Methodist Church', but pointed out that it was 'a precious means of grace woven into the very warp and woof of the fabric of our Methodism'.[48] But by 1960 a Methodist historian was lamenting that it was 'often senile and decaying'.[49] This was perhaps predictable as the old intimacies died out. Another distinctive institution which died out even earlier was the love feast, when the whole society was supposed to meet, quarterly, for a re-enactment of the early Christian agape, usually conducted with biscuits and water, later with tea and biscuits.[50] By the 1930s it was practically dead, or transmuted into the tea and biscuits meetings which remained a feature of Methodist 'sociality'.

In 1900 there were 176 active Methodist ministers in Ireland, and 37 retired. They were under-educated compared with the Anglicans and Presbyterians; only 15% were graduates. This is perhaps surprising in that the Methodist Church controlled two leading secondary schools, Wesley College in Dublin, and the Methodist College in Belfast. The impetus to start these had been partly the system of ministerial itinerancy, which led to the frequent disruption of a child's education. Both schools had large boarding departments. Of the 258 ministers who died between 1880 and 1930, 227 had been born and brought up in the country, and it was still possible in 1930 to claim that 'the chief recruiting ground for our Ministry has been in the rural areas on Ireland'.[51] The itinerancy system however ensured that Methodist ministers never developed the tradition of 'rootedness' which is evident in much of the Presbyterian ministry in that period. Before the turn of the century ministerial training was sketchy. Some candidates were taught at Methodist College, where training of ordinands began in 1870. Some few each year were sent to colleges under the auspices of the English Conference. Most however were untrained, at least in that formal sense, though they were obliged, during their probationary years in their first appointments, to undertake an extensive programme of reading, John Wesley himself having been a keen proponent of a reading ministry. It was not until after the World War I that steps were taken to establish a theological college in Belfast: a building near the university was acquired in 1919, and within a decade had become Edgehill College, where almost every subsequent Methodist ordinand was trained.

Methodism was a centralised church, ever since John Wesley had exercised a strict personal control over his societies, but it was not so centralised on a city, as the Anglicans were on Dublin or the Presbyterians were on Belfast. It was however centred on the annual Conference which alternated between those cities, but every third year met in Cork (or, after 1947, in Portadown). This meeting of the clergy, with from 1877 lay representatives as well, was an easygoing affair as compared with the General Synod or General Assembly, reflecting that 'sociality' on which they prided themselves. But it carried great weight nonetheless. 'It is amazing what influence anyone who bears the stamp of the Methodist Conference has over the people at large', one young minister noted with surprise: 'I find myself an authority on any matter.'[52] And of course because ministers were 'stationed' by conference, with up to a third being moved each year, it impinged very directly on the lives of clergy and their congregations.

The Methodist Church differed from its giant neighbours in that it seems to have had fewer 'nominal' members. This was a characteristic shared with the smaller denominations which were in a sense 'gathered churches', where believers gathered and which have been described by Troeltsch as 'sect type'. By this he meant a 'free union of stronger and more conscious Christians who join together as the truly reborn, separate themselves from the world, remain limited to a small circle, emphasise the law instead of grace and in their circle set up love as the Christian order of life'.[53] One such sect or small denomination was the Plymouth Brethren, which – despite its subsequent name – grew up in Dublin in the 1820s, mainly under the influence of a dissident Church of Ireland minister, J.N. Darby, and other 'evangelical malcontents'.[54] His extensive travels in North America led to its secure establishment there, though in its city of origin, by 1901, it could only claim 133 adherents, and in Belfast 386. Like many smaller churches, with sect type characteristics, and an emphasis on biblical literalism, separatedness and eschatology, it had a tendency to fragment, as stricter believers separated themselves off from the less strict. One Belfast boy remembered its activities in the 1930s, 'cankered' by division between 'Open' and 'Closed' Brethren. While the closed Brethren kept themselves to themselves, and pursued their private and somewhat doleful pilgrimage, the Open Brethren, 'banners flying and sandwich-boards donned like armour' conducted open-air parades and services.[55] The Salvation Army, which first set up its Irish campaign in Sandy Row, Belfast, in 1880, provided the attraction of bands and marches and uniforms, though perhaps surprisingly never grew significantly, in Belfast or beyond. Belfast, however, was a rich breeding ground for groups like these. Especially from the turn of the century onwards gospel halls and missions sprang up, established by Elim Pentecostals, Assemblies of God, Four Square Gospellers, and so on. They provided for some an alternative, for others an addition, to membership of the bigger churches. 'To those of us who were accustomed to the rigid ritual of our various churches', as one man reminisced, 'there was a striking informality about the gospel service.'[56] A different kind of sect, with a different appeal, was Christian Science, which was founded in 1879 in Boston and started work in Dublin in 1902, the Belfast branch opening in 1904.[57] Most Christians in Ireland would have regarded its beliefs as non-Christian. Many of the sects, with their emphasis on particular aspects of Christianity, and their neglect of others, might be regarded as being very near the boundaries of orthodoxy. Christian Science falls outside those boundaries. A church which might be said to be just on the boundary is the Non-Subscribing Presbyterian Church which had its origins in the doctrinal conflicts within Presbyterianism early in the nineteenth century, and which was in effect the Unitarian Church in Ireland. While unitarianism cannot, by definition, be a part of Trinitarian Christianity, the Irish Church was far less removed from traditional Trinitarian Presbyterianism than its counterpart in England. That it retained 'Presbyterian' in its name underlines that traditionalism.

Some of the other larger mainland denominations also had a small presence in Ireland, notably the Baptists. They had been affiliated to the Baptist Union of Great Britain and Ireland since the middle of the nineteenth century, but the 1888 'Down Grade Controversy' in England had its effect on Ireland. So called because of the perceived threat of a down-grading of the preaching of the Gospel, it led to the most noted Baptist preacher of the day, Charles Haddon Spurgeon, leaving the Union in protest against liberal tendencies. The Irish Baptists too reviewed their position, and in 1895 formed their own Baptist Union of Ireland.[58] Symbolically, from that time on their clergy seem to have called themselves 'Pastor' or plain 'Mister'; very few used the title 'Reverend' as they had previously done. Consideration was given to re-affiliating with the British Union early in the twentieth century, but only a few Baptist churches were 'liberal' enough for that, and they maintained their own links with the parent body in England, while remaining affiliated also to the Baptist Union of Ireland. Although there were only a few thousand Baptists in the country, mainly concentrated in Belfast, they had their own training college for ministers in Dublin, in Rockefeller House (built with the aid of American dollars) which was also a school for boys, and a YMCA centre. The 'Irish Baptist Training Institute' maintained a semi-detached relationship with the Baptist Union from its founding in 1892 until it was taken over officially by the church in 1962, and reopened in Belfast in 1963 as a theological college.[59] The subsequent careers of many of those trained in Dublin illustrates the volatility in the smaller denominations. Of the 74 men who entered the Institute between its founding and the outbreak of World War II, seven ended up as clergy in other churches: three as Presbyterians, two as Anglicans, one as a Presbyterian and one as an independent. This 9.5% seepage to other churches might be contrasted with the Anglican experience. Of the 1,172 clergy who were ordained into or served in the Diocese of Connor in the twentieth century, a mere five left the Church of Ireland ministry – one to be a Roman Catholic priest, one a Baptist minister, one who tried his hand at Free Presbyterianism before returning to the Anglican fold (as a layman), one who was deprived of his orders, and one who having been 'inhibited' went to serve in the less restrictive atmosphere of the Church of England. While not strictly comparable with the Baptist statistics, a seepage of less than 1% from Anglicanism is a striking testimony to the difference between a large, formal institutional church and the tenser, more doctrinally edgy atmosphere of a smaller 'gathered' church.

Other small churches abounded. Some 500 people in Belfast in 1901 were members of the Moravian Church, which originated in Bohemia after the martyrdom of John Hus in 1415. Moravians retained an episcopal order, and regarded themselves as a group within Lutheranism. Despite their pietistic leanings, for a spell in the eighteenth century they became a missionary church – and indeed had an influence on John Wesley. By 1900 most Moravians lived in America, usually in rural settlements, and in Ireland they established one in the form of a model village at Gracehill. At their annual conference there in 1913, the minister of Cliftonville church lamented their inability to grow in numbers, giving as reasons 'our foreign name, our comparative isolation, [and] the patent fact that our work here is not supported by a very strong and active organisation behind ...'.[60] There were other churches which also lacked such support. The Reformed Presbyterians, who traced their roots back to the Covenanters, were a small, conservative group, larger than the Moravians but like them not much given to proselytism. Further new Presbyterian churches would arise in the twentieth century. In 1927 an Irish Evangelical Church was formed, a breakaway from the parent Presbyterian body as a result of the latter's perceived doctrinal liberalism. The Free Presbyterian Church, founded by Ian Paisley after World War II, had no links with the Presbyterian Church in Ireland, except in Paisley's choice of a name for his new denomination. His church would disprove the Moravian assertion that 'it is difficult to wrest members from the established denominations'.[61] A well established denomination similar in doctrine to the Presbyterians was the Congregational Church. Indeed such was the relationship between the two that they combined in England in 1972 to create the United Reformed Church. Not so in Ireland where the small Congregational Union retained its own identity. In 1900 it had some 9,000 members throughout the country, though concentrated in Belfast, and with links to the still powerful Congregational Union in England. Those links are well symbolised in the dominating figure of the Revd James Ervine, who kept close contacts with Britain (where he served as a minister for a short time) by his directorship of the London Missionary Society, and his membership of the International Council of Congregationalists, of the English Chapel Building Society and of the British Society for the Propagation of the Gospel among the Jews. In addition he was President of the Irish Congregational Temperance Society, a member of many of the Committees of the Irish Congregational Union, and its Chairman in 1884 and 1889. He was invited to take the chair again in 1911, but declined, as by then he had been retired from active work since 1897.[62]

A religious group which was much smaller, but perhaps more influential, than the Congregationalists did not call itself a church at all. The Religious Society of Friends (Quakers), founded in England by George Fox in the seventeenth century, had no formal ministry, and no sacraments. Their doctrines (especially in Ireland) tended to be orthodox, and there were Quakers who attended other churches as well, becoming thereby rather like the early Methodists, who had been a 'society' rather than a church. The Quaker traditions of philanthropy and pacifism were significant in the history of Ireland, despite the smallness of their numbers.[63] At the Yearly Meeting held in Dublin in June 1899, it was reported that the Society had a total membership throughout Ireland of 2,586. Regular attenders were however about a fifth of that number, and those mostly in Ulster.[64] Attendance at meeting was a different experience from going to church. A Quaker meeting was a time for silence and contemplation; a Dublin meeting in 1901 was described as 'a time of true worship and communion, during which many voices were heard in prayer, and earnest addresses were delivered ...'.[65] One reason for their static or declining numbers was, one of them claimed, 'the aloofness of Friends themselves'.[66] Socially they tended to be fairly exclusive. 'Friends were landowners, millers, farmers and shopkeepers. Some were very wealthy', writes the Quaker historian Isabel Grubb.[67] Her own family saw another side of that exclusiveness when in the mid-nineteenth century Richard and Maria Grubb were 'disowned' (or 'excommunicated') for having 'introduced and encouraged the practices of Music and Dancing in their house'.[68] They were amused to note in later years that many Quakers in Ulster had introduced piano music and singing into their worship as a result of the Moody and Sankey missions in Belfast. Those missions in fact did much to invigorate the common faith held by Irish Protestants.

2
The Churches and Politics, 1900–1922

For 'happy homes', for 'altars free', we grasp the ready sword –
For Freedom, truth, and for our God's unmutilated word.
These, these the war cry of our march, our hope the Lord on high:
So put your trust in God, my boys and keep your powder dry.

Anonymous, 1912[1]

Home rule and Rome rule

The Protestant churches in Ireland were unionist: they maintained that the century-old Act of Union between Britain and Ireland must remain on the statute book. The main churches had, as institutions, opposed Gladstone's first and second home rule bills in 1886 and 1893. The overwhelming majority of Protestant clergy and laity held to the belief that home rule would be 'disastrous'. It was possible for them to summon all sorts of reasons for this: there were constitutional, imperial, strategic, economic and social as well as religious objections to the plan to set up a home rule parliament in Dublin. As the new century dawned, however, the home rule issue was somewhat in abeyance. Other matters claimed the attention of the churches: education was a high priority. The Conservative government's policy – dubbed 'killing home rule with kindness' – had since the mid-1890s meant a concentration by Dublin Castle on economic and social issues, including education. George Wyndham, the Conservative Chief Secretary for Ireland, summarised the position in 1902 when he wrote: 'I shall pass a Land Bill, reconstruct the Agriculture Department and Congested Districts Board, stimulate fishing and Horse-breeding, and revolutionize education. Then' – as he put it with obvious feeling – 'I shall say "nunc dimittis" and let someone else have a turn.'[2] A Liberal successor of his, Augustine Birrell, was announcing five years later with unwonted complacency that 'Ireland is in a more peaceful condition than she has been for the last six hundred years.'[3]

In fact, the first dozen years of the new century saw a continued growth of sectarianism, helping to mould attitudes and create the embittered atmosphere in which, when it arose again, the home rule question would be debated even more hotly than before. In 1901 a series of lecture tours throughout Ulster was undertaken by M.J.F. McCarthy, a Cork Roman Catholic who was a barrister and a graduate of TCD. His first book, *Five Years in Ireland*, published in March 1901, reached its tenth edition by November 1903, and his *Priests and People in Ireland*, published in 1902, was in its fifth edition by 1905. Other books followed, all on the same theme: 'that Priestcraft is omnipresent, all-pervading, all-dominating' and that 'sacerdotal interference and domination in Catholic Ireland ... will be found to be the true and universal cause of that universal degeneracy upon which we so commiserate ourselves'.[4] These books furnished Protestants with ammunition to attack the influence of the Catholic Church. They stoked up Protestant fears, with stories of how the rector of St Luke's in Dublin was 'booed and hissed by a crowd';[5] of how a minister had 'bloody offal' thrown at him;[6] of the violent boycotting of the Irish Church Mission's Limerick dispensary, and the personal abuse suffered by its superintendent, Dr Long.[7] McCarthy's books found their way onto thousands of Protestant bookshelves. His lectures were widely reported and hugely attended, as he attacked what he saw as the pernicious influence of a resurgent Catholicism. Certainly the Irish Catholic Church seemed triumphalist in the new century. Home rule might no longer seem to be a live issue, but the church had not been slow to fill the vacuum. The '*de facto* Irish state' of these years was, as Emmet Larkin has put it, 'essentially a confessional one'.[8] As one Protestant nationalist complained to another, 'politics have now been, in fact, like everything else, absorbed into the Church'.[9] New organisations had sprung up for Catholics. The United Irish League, founded in 1898, the Ancient Order of Hibernians, which saw phenomenal growth in Ulster in the years 1905–9, and the sectarian Catholic Association founded in 1902, all added to Protestant unease, as did the official actions and statements of the Catholic Church. In

February 1901 it prohibited the faithful – even those of Privy Councillor rank – from attending Memorial services for the late Queen, thus reinforcing the Protestant conviction that Roman Catholicism was fundamentally 'disloyal'.[10] In June 1901 a Curial directive ruled that Protestants ('heretics') could not be godparents to a Catholic child.[11] In that same month Belfast Catholics held a Corpus Christi service, preceded by an open-air procession of 15,000 people, who marched 'silently, without bands, and with furled banners'. It seemed to many an ordinary Protestant that 'the Catholic priest had taken over his streets'.[12] When, in November 1903, TCD held out an olive branch by inviting Cardinal Logue to enter discussions for the provision of Catholic chaplains and catechetical instruction in the College, the response was terse in the extreme: 'I beg to acknowledge the receipt of your letter of yesterday's date, and to state that I can be no party to the arrangements proposed therein.'[13] All these manifestations of Catholic triumphalism and militancy looked the more sinister to Protestants when set against the political scene, even though the Conservatives were in power until 1905. The policy of 'killing home rule with kindness' was interpreted as the 'Romanisation of Ireland by the Balfour brothers'.[14] The Presbyterian newspaper in 1902 stated baldly that 'Home Rule has not been killed by kindness. We have pampered the lion's cub with the children's bread, until the evil beast has grown upon our hands and is already whetting its appetite upon the children themselves.'[15] When in 1904 the civil servants in Dublin Castle came up with a plan for devolution – a mild version of home rule – it seemed that Protestants were not safe even under a Unionist government; the affair 'restored Protestant suspicions to the fever-pitch of 1886'.[16] But it was four years later – and by then a Liberal government was in power – that there arose a crisis which had far wider and longer-lasting effects.

The Papal decree, *Ne temere*, issued in 1908, dealt with marriages in which one of the partners was a non-Catholic, and laid down regulations regarding the bringing up of children of any such marriage in the Catholic faith. It was condemned in the Church of Ireland General Synod of 1908, and by all the other Protestant churches thereafter, but it took the McCann case of 1910 to create a *cause célèbre*. The Revd William Corkey, minister of Townsend Street Presbyterian Church in Belfast, brought to light in November 1910 the case of Mrs Agnes McCann, a Presbyterian who had married a Roman Catholic. Their marriage was happy, and as agreed they each attended their own church. There were two children. Then (it was claimed) Mr McCann's priest visited the house to tell the couple that their marriage was invalid in the light of the *Ne temere* decree. Mrs McCann refused to remarry in a Catholic ceremony. Her husband allegedly began to ill-treat her, and in the end vanished, along with their two children and their furniture, leaving Mrs McCann destitute. She had turned to her own minister, Mr Corkey, who set about publicising the 'scandal', and in a lecture to the Knox Club in Edinburgh, issued as a pamphlet, he spelled out the dire consequences of *Ne temere*. It was 'a danger to the Commonwealth, because it strikes at the home'; it challenged 'the supremacy of British Law'; it meant 'that the Church of Rome can absolve a man from the solemn marriage vow'; it showed 'that Rome is still prepared to inflict cruel punishment on any members of the Protestant Church over whom she gets any power'; it was 'a proselytising instrument to help in the reconquest of Britain for Rome'.[17] The Revd J.B. Armour, the Presbyterian home ruler, was not impressed: 'The general opinion among those who know the woman is that she is no great shakes and to use her case for purely political purposes shows the straits to which the Ulster Tories are reduced.'[18] Even ordinary anti-home rule Protestant clergy were uneasy. The Revd F.E. Harte recorded that he felt uncomfortable while speaking against the decree at the Grosvenor Hall because 'there was rather too much of a political atmosphere'.[19] Certainly while the details of the case are in doubt there can be no doubt at all of its effectiveness as a rallying cry for Irish Protestants. It saw the Protestant churches acting together in a way that foreshadows their political unanimity a year later. And in January 1912 the Ulster Women's Unionist Council launched a petition against the decree which within a month had collected the signatures of 104,301 women,[20] foreshadowing the signatures of their husbands and brothers and sons on Ulster's Solemn League and Covenant nine months later.

The McCann case created what has been called a 'moral panic' among Protestants.[21] It gave a focus to all their frustrations and fears, and its timing was perfect. William Corkey delivered his lecture in Edinburgh on 21 February 1911 – the day on which Asquith introduced the Parliament Bill into the House of Commons. The general elections of 1910 had left the Liberal government dependant upon the Irish Nationalists' votes, and the Parliament Act would soon remove the Lords' veto, upon which unionists had been able to rely in 1893. The election in January 1910 was preceded by a manifesto 'To the Electors of Great Britain' by eleven ex-Moderators of the General Assembly, declaring their unswerving commitment to the union; 'no more important manifesto' appeared during the election campaign, the Presbyterian newspaper declared.[22] From an Anglican point of view, it was significant that 'the

Protestants of Ireland have stood together in the present political crisis'.[23] Surveying the results of the December election, R. Dawson Bates, secretary of the Unionist Associations of Ireland, reported that many of the favourable results in Britain were 'mainly due to the distribution of leaflets dealing with the attitude of the Church of Rome towards the Protestant religion in Ireland, and to the influential advocacy of our Cause by the Presbyterian and Methodist Ministers who represented us ...'.[24] Their advocacy had not been sufficient however to dislodge the Liberals from government.

The passing of the Parliament Act in August 1911 was the signal for renewed Unionist activity in Ulster, directed by Edward Carson, the Dubliner who since February 1910 had been leader of the Irish Unionists. Carson wanted to be sure that 'the people over there really mean to resist', as he wrote to James Craig.[25] A great demonstration at Craig's family home, Craigavon, with 50,000 men on parade, gave sufficient proof. It was this demonstration which, as A.T.Q. Stewart puts it, 'determined both the policy and the methods to be followed during the next three years'.[26] For the churches, the determination to resist home rule was only deepened by a further papal intervention, in October, when the motu proprio *Quantavis diligentia* seemed to suggest that the Roman Church was now claiming that its clergy should be immune from 'process of civil or criminal law of every sort in any lay court'.[27] Coincidentally, just over a week after its publication – though before there had been any reaction in the press – Church of Ireland and Presbyterian representatives met for the first time in bilateral talks to discuss matters of mutual interest. High on the agenda was the *Ne temere* decree. Such ecumenical contacts and co-operation would increase greatly over the subsequent two years of political ferment.

The Presbyterians were the first to mobilise against the new home rule threat, with a Convention on 1 February 1912, chaired by Thomas Sinclair, the former Liberal and a leading Presbyterian layman, and attended by 'almost half the adult male Presbyterian population in Ulster'.[28] Resolutions were passed, with religious themes prominent: that 'our civil and religious liberties would be greatly imperilled'; that 'the philanthropic and missionary enterprise of our Church at home and abroad would ... be greatly curtailed'; that 'Presbyterian minorities in all parts of Ireland (many of them consisting of settlers from Scotland)' would be in danger; that children would be at risk from the 'denominationalising' of education by the Roman Catholic hierarchy. Mammon as well as God got a mention in that 'industrial and agricultural interests would be seriously crippled'.[29] The reference to Scottish settlers was significant. During the 1893 home rule crisis, the Irish Moderator had attended the Church of Scotland's annual meeting, chiding his Scots brethren for their 'hollow expressions of sympathy', and emphasising Irish fears at the 'prospect of being handed over to the tender mercies of a Parliament dominated by Romish priests'. He got short shrift from Lord Balfour of Burleigh who said he had heard too much about 'the Roman Catholic Church'.[30] This time, the efforts to elicit support from Scotland paid off.[31] It was noteworthy too that the Presbyterians made the point that they were appealing to a government 'with whose policy, apart from the question of Home Rule, so many of us are in general sympathy'.[32] But for most Presbyterians, this home rule issue was 'not a political question', for as the Revd Robert Barron put it, 'in ordinary political questions, I took no part and expressed no opinion'.[33] This issue was different. 'In those days', he recalled, 'I often spoke and preached on the subject, and I urged the people to resist even to death.'[34] The Presbyterian newspaper felt the same, and believed that 'history will justify Protestant resistance to the present attempt – humiliating on the part of England, hypocritical on the part of Rome – to establish Roman ascendancy in Ireland'.[35] And yet another affair seemed to underline the risks Protestants faced. In July, the same Robert Barron organised a Sunday school outing which suffered what became known as the 'Castledawson outrage', when some 50 pupils and teachers were 'attacked and battered' by a party of some 250 men 'wearing green hats'.[36] The Belfast Presbytery speedily appointed a commission to 'present a statement of their case to their Brethren in England and Scotland'.[37] It was an extraordinarily timely illustration of how home rule would mean Rome rule. The Methodists meanwhile had followed the Presbyterians with their own anti-Home Rule Convention in March at the Ulster Hall. They were anxious to 'disavow, as utterly alien to the spirit which we have inherited ... all feelings of enmity or ill-will to any class in the community'.[38] They reiterated their Conference's anti-home rule resolutions of 1886 and 1893, adding in addition the anxieties caused by *Ne temere*. And in a flourish characteristic of this small, evangelical and puritan church, they emphasised their fear of 'very wide and far-reaching evil consequences, especially imperilling the best interests of education, temperance and Sabbath observance'.

All the Protestant churches were officially represented at an Easter Tuesday parade and service at Balmoral (outside Belfast), attended by the recently appointed leader of the Conservative party Andrew Bonar Law, who himself boasted Presbyterian Scots and Ulster forebears. In an ecumenical united front, the service was conducted by the Presbyterian

Moderator and the Church of Ireland Primate. The numbers were double those who had paraded before Carson at Craigavon, and in addition, the proceedings were attended by 70 MPs from Britain. The demonstration had its effect. When Bonar Law got back to London he revealed to a friend just how much the experience had meant to him: 'The demonstration last week was a great surprise to me. It may seem strange to you and me, but it is a religious question. These people are in serious earnest. They are prepared to die for their convictions.'[39] Three months later, Law attended another demonstration, this time at Blenheim. It was there that he made his bold statement: 'I can imagine no length of resistance to which Ulster can go in which I should not be prepared to support them, and in which, in my belief, they would not be supported by the overwhelming majority of the British people.'[40]

In the months after the Balmoral demonstration, further church pronouncements were forthcoming. In April, the General Synod of the Church of Ireland met in Dublin in a special session. It opened with the singing of the hymn that was rapidly becoming the national anthem of the Irish Protestants, 'O God our help in ages past'. The Primate, J.B. Crozier (who had been Bishop of Down until Archbishop Alexander's resignation in the previous year) opened his address by stating: 'Our Church knows no politics, and our Church is tied to no political party.' As to the question of home rule, however, he noted that 'the vast majority of the member of all the Protestant religious bodies in Ireland are absolutely united'. He dwelt upon the threat to the unity of the Empire and the likelihood of 'anarchy and civil strife' should home rule be enacted. Although there was no specific mention of 'Rome rule', Crozier praised the Ulstermen, noting their resolve 'to resist extra-constitutional measures by extra-constitutional means'. The body of 402 synodsmen then voted for a resolution reaffirming 'constant allegiance to the Throne, and our unswerving attachment to the Legislative Union ...'. It was passed with only five votes against. Further resolutions of a similar nature were passed, and the proceedings ended with the singing of the national anthem.[41] One of the few dissentients was the Revd J.O. Hannay, whose claims then and later that '[T]he resolutions of the Synod are not "The Voice of the Church"' were dismissed by the 'sturdy Northerner' Archdeacon Pooler as a 'rhodomontade of nonsense'.[42]

In June 1912, even the Reformed Presbyterian Church, which was strictly non-political to the extent of advising its members to abstain from voting at elections, passed a resolution condemning home rule, 'because the measure transcends ordinary politics'.[43] The statements and utterances from pulpit and platform throughout that summer of 1912 were indeed hardly the stuff of normal political discourse. 'Their cause was just. The unchanging God was on their side', declared the Methodist minister, L.W. Crooks.[44] 'They were prepared to lay down their lives in defence of their principles', said the Rector of Newtownards, the Revd W.L.T. Whatham.[45] 'It is not sufficiently realised', said Bishop D'Arcy of Down, 'that behind Ulster's opposition to Home Rule there is an immensely strong conviction which is essentially religious. We contend for life, for civil liberty, for progress, for our rightful heritage of British citizenship. We also contend for faith and for the freedom of our souls.'[46] But the greatest and most publicised proclamation of such views came on Ulster Day, 28 September 1912. A trio of clergy, the Dean of Belfast, the Presbyterian Moderator, and the Methodist Vice-President, formed a subcommittee to help with its planning.[47] Their input was significant for, when the day came, its religious overtones were evident to all, even though the Archbishop of Dublin vainly hoped to 'avoid identifying the Church generally with the proceedings'.[48]

Looking back on it, St John Ervine caught the religious mood, when 'Belfast suspended all its labours and became like a place of prayer. Very solemnly, the people went to church to prepare their minds for their responsibilities, as postulants prepare themselves for consecration.' It was to be a day of dedication, when men and women sang 'O God our help in ages past' 'like a hymn of battle'.[49] A.F. Moody, later to become Presbyterian Moderator, recalled 'the signing of a covenant with a fervour equal to that of the old Scottish Covenanters at Greyfrairs'.[50] Looking forward to the day, an Ulster newspaper noted that it would be 'solemnised by prayer and hallowed by the sanctities of religion'.[51] Looking back on it *The Times* was even more struck by the Covenant as 'a mystical affirmation ... Ulster seemed to enter into an offensive and defensive alliance with Deity.'[52] The Covenant was signed by nearly half a million men who, 'humbly relying on the God whom our fathers in days of stress and trial confidently trusted', pledged themselves to 'defeat the present conspiracy to set up a Home Rule parliament in Ireland'.

A year later, preaching in Belfast Cathedral on the anniversary of 'Ulster Day', Bishop D'Arcy surveyed the events of past and present: 'The steps which have been taken of late are of the most awful import. They have been taken, we believe, in the path of plain duty – in defence of things for which a man must be prepared, if need be, to make the utmost sacrifice.'[53] The signing of the Covenant had been followed by the establishment of the Ulster Volunteer Force (UVF), soon to be armed with guns brought in from Germany. Sir Edward Carson and his followers were preparing a Provisional government to take over Ulster in the event of the passing of home rule. There was talk at Westminster of the possibility of an exclusion clause to allow Ulster an exemption, temporary or otherwise, from the provisions of the Home Rule Bill. Lest the focus become too political, the Revd T.C. Hammond wrote an article for *The Nineteenth Century* in February 1913. It was devoted to showing his readers 'that the solid mass of Irish Protestants of all denominations and of every class in life regard the religious question in its extensive ramifications as the one issue of paramount importance'.[54]

Education issues

Norman Sykes gave the Wiles Lectures at Queen's University, Belfast in 1959. He spoke of 'the question of education as the nerve-centre of the present-day relationship of Church and State'.[55] As it happens, he was not speaking of Ireland, but he could well have been. Education was and has remained one of the thorniest issues with which successive governments – in London, Belfast and Dublin – have had to deal. It is true the world over, as it must be: who is to control the education which the young receive, for whoever determines their early education has enormous influence on the citizens of tomorrow? Should that influence be under the control of the state, of the church (or churches), or of local people? The government had instituted in Ireland – back in 1831 – a system of 'national' schools at primary level which were to be non-denominational. Decades before tackling the question of primary education in England, the decision was taken that Ireland's future citizens would be educated within a secular – or at any rate non-sectarian – education system. It had not worked. The 'national schools' had become – effectively – denominational schools, depending upon which religious denomination set them up or dominated in the locality. The system had exposed fundamental divisions of opinion in Ireland. There was no strong lobby, such as was to emerge in England in the second half of the nineteenth century, for a purely secular system of education. The question in Ireland turned on how religious and moral values were to be transmitted within the system. For Roman Catholics, the answer was simple: schools or colleges must teach the faith. The guardian of that faith was the church, and therefore educational institutions must come under the control of the church, and of its clergy. Bishop Conroy stated their case, in an historically inaccurate sermon which nevertheless illustrates how Catholics had come to view the National System: '... the Nonconformists in England and the Presbyterians in Ireland and Scotland have conspired, in language bristling with insult and menace, to force upon the Government a system of godless education for Ireland, not on the ground that it alone is just, or because it is plainly best for the country, but avowedly because it is pernicious to Catholic interests ...'.[56] For Protestants, the answer was less clear cut. They, like the Roman Catholic Church, wanted schools and colleges to teach religious and moral values. But for many Protestants, this could mean 'simple Bible teaching', which need not be given by the clergy, nor need it be denominational. It is true that the Church of Ireland, like the Church of England, had aspired to be the nation's schoolmaster, and its stance therefore tended to be a shade nearer that of the Roman Catholic Church; hence Conroy's specific naming of the Presbyterians, who, with the smaller Protestant denominations, had been historically and financially less well placed to control education, and so tended to look askance at the pretensions both of the Roman Church, and the Church of Ireland. But given their minority status, especially after disestablishment, Anglicans could hardly claim either the right or the ability to educate the nation's children. At one level, however, the Church of Ireland did have a strong position, in its ancient hold on Trinity College, Dublin, the best-endowed and most respected institution of higher learning in Ireland.

At the beginning of the twentieth century, it was the 'university question' which was uppermost in people's minds. It had been jangling the nerves both of the churches and of the state throughout the nineteenth century. Until 1845 the sole provider of university education in Ireland had been the University of Dublin, known more popularly by the name of its sole constituent college, Trinity College. Since the sixteenth century TCD had been Ireland's equivalent of Oxford and Cambridge, with the same kind of Anglican foundation. Peel's Conservative government in 1845 had initiated the foundation of non-denominational colleges at Belfast, Cork and Galway, but their deliberately non-sectarian nature had soon brought them into disfavour with the Roman Catholic hierarchy, who dubbed them 'godless colleges'. Their answer in 1854 was to establish a Catholic College at Stephen's Green, in Dublin, with John Henry Newman as its first Rector. It received no state endowment, though its students could hold scholarships from and gain degrees through the Royal University of Ireland, an examining body set up in 1879 in an attempt to meet Roman Catholic demands. The RUI became the degree awarding body for all non-Trinity students in Ireland – including those at the 'godless colleges' and at Magee College. Successive Liberal and Conservative governments in the nineteenth century were confronted by the 'university question', and found themselves under attack from all sides. Catholics wanted state funding for their Stephen's Green College, perhaps through a carve-up of Trinity's endowments. Nonconformists, in England and in Ireland, found the possibility of state monies going to such an explicitly Catholic institution hard to stomach. Presbyterians, like Catholics, cast longing eyes on the wealth of Trinity. Anglicans – and especially the Anglican clergy – were devoted to TCD, which housed their Divinity School, and most were not unsympathetic to the idea of state funds for Catholic university education – certainly if it meant that Trinity would be left alone. So the Protestant churches themselves were in disarray over the question of how to proceed.

A.J. Balfour, whose policy of 'killing home rule with kindness' was to cause such concern to Irish Protestants around the turn of the century, had declared with some realism in 1889 that 'from my own experience of Ireland ... you cannot altogether escape sectarian education', because, as he observed, 'the differences are so deeply ingrained in the very fabric and substance of society they cannot be ignored'.[57] He was right, but was to discover that any attempt to produce a final solution of the university question was to stir up sectarian animosities and to reveal deep divisions within Protestantism. Churches which had spoken with one voice in 1893 against the second Home Rule Bill soon found themselves at odds over the university question.

At the beginning of 1897, the home rule MP for Kildare North, Charles Egleton, tabled an amendment to the Queen's Speech, claiming that 'it is the duty of the Government immediately to propose legislation with a view to placing Irish Catholics on a footing of equality with their fellow countrymen in all matters concerning University Education'.[58] The predictable response from the Unionist Orangeman who represented Belfast South – William Johnston – was to deny the reality of Catholic grievances, and he 'implored the Government not to be led away by the hope of conciliating the Irish party'.[59] The Presbyterians were watching developments carefully, and at the General Assembly in June the outgoing Moderator, the Revd H.M. Williamson, warned that non-sectarian education was under attack by 'the combination of the representatives of Conservatives and Liberals, of Trinity College and the Vatican'.[60] This was a powerful alliance of interests so the Presbyterians ought to demand from the government that 'if against our protest and opposition you do grant a University to Roman Catholics, then we demand that you establish another University for us in which the Presbyterian Church shall have all the privileges which the Protestant and Episcopal Church has in Trinity College, and which the Roman Catholics will have in their

University'. This proposal was greeted enthusiastically by the nationalist *Freeman's Journal*, though it added, tongue in cheek, but with some accuracy: 'It does not do, however, to abandon candidly the fiction that Presbyterianism is the foe of denominational education.'[61] Edward Carson, one of the MPs for Dublin University, rose in the Commons that same day to support 'a university ... which Roman Catholics might not only attend, but which should harmonise generally with the views of Roman Catholics, and which should have its doors as widely open to members of all denominations as Trinity College, Dublin'.[62] There seemed to be emerging, as one observer put it in 1898, 'a closer proximation of Roman Catholic to Protestant opinion than formerly prevailed, [and] an eagerness on both sides of the controversy to find a common ground of agreement'.[63] Finding the right compromise, acceptable to all sides, was not going to be easy, and Balfour asked the Liberal R.B. Haldane to go to Ireland, interview all interested parties, and come back with recommendations. The Haldane plan was circulated to the cabinet in the autumn of 1898.[64] It was the sort of payoff the Moderator had suggested. Trinity was to remain untouched; there was to be a university in Belfast and another in Dublin. These would 'have both a teaching (internal) and examining (external) side. In this way those Catholics and Protestants, in whatever part of Ireland, who were not able to attend the lectures of recognised teachers, would be able to present themselves for external degrees.' Haldane reported that there was general approval for such a scheme, although 'ten to twelve of the Ulster members would find it impracticable to vote in its favour, on account of the Orange element in their constituencies'. Balfour, in a memo circulated with the Haldane report, underlined the implicitly 'concurrent endowment' aspects of the plan, emphasising that 'what is done for the Roman Catholics in Dublin should, <u>in the same bill</u>, be done for the Presbyterians in Belfast'.[65] While the plan went into cold-storage as cabinet energies were diverted to the more pressing needs of the war in South Africa, it was already under fire. Salmon, the Provost of Trinity, wrote to Edward Carson to complain that 'you seem to me not to apprehend the danger of ruin this University is in'.[66] He and Bernard, fellow and future provost, were loud in their denunciations of a scheme which, it seemed to them, would undermine their *alma mater*. While Trinity was to be left untouched, it would face unfair competition from two new and well endowed institutions. This stand provoked a reaction among their fellow Protestants. The Presbyterians began to call for 'the nationalisation of Trinity College ... as the true solution to the university question in Ireland'.[67] Rather than accept the Balfour–Haldane plan, they now declared that no plan which excluded Trinity would be acceptable. The Methodists, who had nothing whatever to gain from the plan, raised a similar cry.[68] The clamour produced no official response, but seemed implicitly to have failed, for when the Robertson Commission was appointed in July 1901 it was charged with inquiry 'into the present condition of the higher, general and technical education available in Ireland outside Trinity College, Dublin'.

The Robertson Commission exposed the fault lines running not only between Catholics and Protestants, but among the Protestants themselves. The Roman Catholic Bishop O'Dwyer of Limerick gave evidence to the commission, proposing university status for Queen's in Belfast and for the Catholic college in Dublin.[69] Hamilton, the President of Queen's, Belfast – and a Presbyterian – wanted to maintain the existing framework, but to incorporate within it well endowed colleges in Belfast and Dublin.[70] Bernard of Trinity agreed with that, reckoning that an endowed Catholic college limited to Catholics would do least damage to Trinity. The Church of Ireland case was put by Bishop Crozier who started off by saying that 'we deplore the possibility of further widening the breach between religious denominations in Ireland', but went on to admit that the only solution appeared to be the endowment of the Roman Catholic college. The Methodists did not agree, though were somewhat divided among themselves. While McIntosh, President of the Methodist College, Belfast, went along with the Catholic college plan (but unlike Bernard he wanted it open to all), the Vice-President of the Methodist Church declared his Church 'perfectly satisfied with existing arrangements'. When one commissioner queried the fairness of Methodists (1% of the population) dictating terms to Catholics (70% of the population) Vice-President Nicholas admitted that Methodists would, at a pinch, acquiesce in the affiliating of a Catholic College to the Royal University, while balking at the endowment of a Catholic University. The Presbyterians were not so easily to be cajoled. In a long submission they argued against the exclusion of Trinity from any settlement. TCD, with 'its beautiful grounds and rich revenues' was still, they maintained, an Anglican stronghold, and as long as that held good, 'Roman Catholics, in common with Presbyterians and others, have the right to feel aggrieved and complain that the principle of religious equality is not impartially carried out'. In a sentence which epitomised this splitting of the Protestant camp they declared: 'It must be admitted that Roman Catholics have this one educational grievance; no other will bear examination.' Here was the old Presbyterian anti-establishment mentality at work. Here too was a grudging admission that Roman

Catholics had a valid grievance. But here too was a growing chink in the non-sectarian armour of the Presbyterians. For Hamilton of Queen's was in favour of the endowment of a Catholic College, as was Leebody, the Presbyterian President of Magee, who – aware of the position of his own institution – advocated 'the open and honest endowment of denominational colleges'.[71]

The Commission's final report revealed, as did the submissions from the various churches, a lack of consensus. Eight of the thirteen commissioners signed statements dissenting – at least in part – from the report's nineteen recommendations. The main complaint was the one the Presbyterians had trumpeted: that Trinity ought not to have been excluded from its terms of reference. This was the position inherited by the new Liberal government, which might just manage a solution to the whole problem by incorporating TCD in a settlement, and thus avoid losing the support of English nonconformists who were even more hotly, and certainly more genuinely, devoted to the non-sectarian principle than their Irish brethren. But as the new Chief Secretary, James Bryce (himself an Ulster Presbyterian and a TCD graduate), was warned: 'Trinity, being the last defence of the ascendancy people, will no doubt give you trouble.'[72] A new Royal Commission appointed by the Liberals was now sitting, charged under Lord Justice Fry with enquiring into the 'present state of Trinity College, Dublin, and the University of Dublin'. The tide of opinion seemed to be running in favour of 'nationalisation' of Trinity. That, Bryce reckoned, would keep everyone happy except 'the representatives of Trinity College and the ultra-Orange faction'.[73] He held his hand until the publication of the Report in January 1907. Like the Robinson report, it was far from unanimous in its recommendations – though on balance Bryce reckoned that it strengthened his hand in terms of the 'nationalisation' approach. By the end of the month, his proposals had been made public. He planned to create a new, enlarged National University of Ireland, consisting of Trinity College, the Catholic College, and the Queen's Colleges of Cork and Belfast. The Presbyterians and Catholics would have accepted the plan; the Anglicans did not. Trinity was determined to defeat the scheme. The *Belfast Newsletter* felt that way too, not so much for love of TCD as from a fear that the plan was but a step towards a fully-fledged Roman Catholic University.[74] There were Presbyterians too who felt much the same thing, but by the time they were airing their views at the General Assembly in June, Bryce had gone off to Washington as British Ambassador, and Augustine Birrell had 'crossed that odious Irish Channel' to become Chief Secretary.[75] He concluded that while a university scheme was urgently needed, his predecessor had been too hasty in making public his plans. The old Balfour–Haldane scheme would be preferable, and 'would only excite the opposition of some of the Orange lodges and that portion of the General Assembly which hates Trinity College, Dublin, even more than it loves itself.[76] He introduced his bill in the Commons in March 1908. The Irish Unionists split: James Craig taunted the Liberals, wanting to know from them, 'especially from the Nonconformists, how, in the face of their opposition to denominational education in this country in the past, they could bring themselves to vote for this bill'.[77] Carson rose immediately after him to welcome the bill. Balfour led the Conservatives – minus the Ulster members – in support of Birrell's rehash of his own plan. A special meeting of the General Assembly convened to condemn the Bill, but only half-heartedly. There were hardly grounds for outright opposition. The Catholics, as they had admitted, did have a grievance. The bill became law. 'The Irish University Question had at last been answered.'[78] The Trinity–Anglican nexus had saved the University of Dublin. The Catholics now had a subsidised institution they could attend with their church's blessing. Only the Presbyterians had been foiled; or had they? The new National University was to be open to all; the Catholics still did not have a fully-fledged Catholic University. Belfast's College had become the Queen's University of Belfast. A decade and more of inter-denominational friction was at an end. The Protestant churches could unite in the face of greater dangers. One, as we saw, was the Home Rule Bill of 1912; the other was the European war.

The impact of war

The chief problem facing the British government in the summer of 1914 was what to do about Ulster. The Home Rule Bill would become law quite soon – the House of Lords no longer being able to exercise a veto. The Curragh Incident in March had erupted because of what Field Marshal Lord Roberts described as 'the wicked plans of Government to quietly assemble troops in Ireland, and then proceed against Ulster as if it were inhabited by a rebellious enemy instead of our own kith and kin, whose sole desire is to be the loyal subjects of the crown ...'.[79] The government in fact backed down from any threat of using force against Ulster, with the result that it 'was neither capable of preventing any declaration of a provisional government in Belfast, nor of responding adequately to the gun running that was the sequel to its paralysis'.[80] There was no official reaction by the Protestant churches to this escalation of the home rule crisis, but individual clergymen acted as they saw fit. The Revd William Corkey – of McCann case fame – 'enrolled as a member of the Ulster Volunteer Force'.[81] Archdeacon Abbott attended a review at Glaslough, in Co. Monaghan of the UVF who were soon to receive a consignment of 128 rifles which were being shipped in from Germany.[82] The Revd John Clarke, an Antrim vicar, used his motor car to collect five bundles of rifles when they landed at Larne on 24 April.[83] The Revd R.G.S. King, Rector of Limavady, published a pamphlet in 1914 in which he proclaimed: 'We are solemnly pledged – cheerfully, willingly pledged – never to submit to this [Home Rule Parliament]. We have called Heaven to witness our determination to resist to the bitter end ... We would not take up arms, did we not believe that the God of Hosts would go forth with our armies.'[84] But Ulstermen were destined to die on a very different battlefield. 'Little did we think', wrote Archdeacon Atkinson describing his blessing of the UVF colours at Lenaderg, 'that in the next year those battalions would be marching in Flanders as part of the Ulster Division in the Great War'.[85] When the war began on 4 August 1914, John Redmond, leader of the Nationalists at Westminster, and Carson, leader of the Unionists, pledged Irish support for the struggle. The home rule crisis had been submerged beneath a greater one. James Richardson of Bessbrook, the notable Quaker businessman, admitted that when, 'like a thunderclap, the lurid cloud hanging over Europe burst, I felt almost relieved'. But then he added, guiltily, '[S]o selfish and petty we are, O purblind race of miserable men!'[86] Others were not so self-questioning. James Winder Good, an Ulsterman who had moved (politically as well as geographically) to Dublin, caustically summed up the attitude of the Ulster Protestants:[87]

> I have heard a clergyman in an address designed to show that at every crisis God had miraculously intervened to save Ulster Unionists, quote as proof of his argument the fact that when the whole world believed Home Rule to be inevitable, Germany marched her forces across the Belgian border. To drench continents in blood in order to save the political prestige of a fraction of the population of an inconsiderable province does not strike the ordinary mind as a miracle of grace, but it is an excellent example of how the theory of a 'Chosen People' works out in practice.

But in many respects, the attitude of Irish Protestant clerics was not so different from that of their English contemporaries. The Bishop of London proclaimed that the nation was 'engaged in a Holy War ... Christ died on Good Friday for Freedom, Honour and Chivalry, and our boys are dying for the same thing ...'.[88] Churchmen were not alone in uttering such sentiments, for 'generals and politicians too liked to conceive of the war in religious terms', but not in denominational ones, for 'on the Western Front, where Protestants, Jews and Catholics fought on both sides, men were asked to believe that they had God on their side'.[89] Dean Grierson, speaking at the Ulster Day anniversary, took up that theme. Germany might be a mainly Protestant country, and Austria a Catholic one, but the war was not being fought 'between different branches of the church of God', but between those who believed in 'justice, liberty and truth' and those who did not. Since it was the anniversary of the Covenant, the Dean added: 'The call of Empire would not drown the call of Ulster, for it was the same call only in louder tones. It was the call of liberty and freedom.'[90] J.H. Bernard, now Bishop of Ossory (and soon to be Archbishop of Dublin), speaking at St Canice's Cathedral, Kilkenny, made the same point, but without Dean Grierson's provincial spin: 'We are fighting for freedom, and for nothing less. For this is the long impending struggle between Liberty and militarism. We are fighting for the free development of free peoples against the rule of the iron hand. We are fighting – I fear we must say it – for civilization against a sudden revival of barbarism.'[91]

It is estimated that some 210,000 Irishmen eventually heeded the call. The UVF became a new 36th (Ulster) Division, composed of more than 26,000 men and 4,000 reservists, which was over a third of the total pre-war membership of the volunteer force.[92] The youth organisations which had flourished before the war also played their part in furnishing men for the army. In July 1911 members of the Armagh Boys' Brigade had marched through the city as an escort for their Unionist MP, J.B. Lonsdale, in celebration of his baronetcy. On Ulster Day 1912 they were matching again, alongside the Anglican Church Lads' Brigade, another organisation with similar ideals.[93] One wonders how many of those boys were now on their way to the western front. Certainly in Dublin, veterans of the Boys' Brigade accounted for nearly half of the Protestant recruits from that city.[94] The Boy Scouts too would 'do their duty to God and the King'. In 1913 the troop from the Royal School, Dungannon, became the first Irish scouts to camp in Europe. On their second visit, in the summer of 1914, they had to abandon their camp near the Swiss border when war broke out.[95] How many of those boys returned to France for sterner purposes in subsequent years?

The war impinged on life throughout Ireland. Nearly a quarter of a million men were serving with the forces. Many clergy were abroad as military chaplains. The national demand for foodstuffs, textiles and shipping meant that the economy thrived; 'indeed it is ironic that the most affluent years of the Union were those immediately prior to its ending'.[96] But in the memory of Irish men and women, the war years would for ever after be associated with blood sacrifice. The Bishop of London had identified the death of young soldiers in the war as a sacrifice comparable to that of Christ on Good Friday. Praising the soldiers at the front, the Revd J.O. Hannay noted their sterling qualities, telling the Archbishop of Dublin in May 1917 that it was 'difficult to account for this general and very beautiful character among our troops except on the hypothesis of a new birth through a baptism of blood'.[97] The Archbishop echoed these sentiments when he preached in St Paul's Cathedral a few months later: 'They have passed, these brave spirits, into the other world, in the very act of sacrificing themselves that others might live larger and better lives. And He who knows all can be trusted to shelter them with His pity.'[98] But blood sacrifice was not to be limited to the troops on the western front. The republicans – or a minority of them – rose in rebellion in Dublin at Easter, 1916. England's difficulty was Ireland's opportunity, or – as most Britons and most Irish Protestants saw it – the Empire was being stabbed in the back at its time of greatest danger. For the rebels, and especially for the poet and rebel leader Patrick Pearse, 'the patriot people was equated with Christ, and hate and violence were sanctified'.[99] The military odds were hopelessly stacked against the insurgents, but perhaps Pearse 'envisaged the rising not as a military operation, but as a deliberate blood sacrifice. The handful of rebels in their hopeless adventure would save the Irish people.'[100] Certainly it was their deaths, by firing squad, which swung public opinion in Ireland behind their cause. Irish Protestants were appalled. The Executive Committee of the Irish Guild of the Church, founded just before the war to promote 'Irish ideals' within the Church of Ireland, met in June and affirmed 'its loyalty to His Majesty the King', and 'being a strictly non-political body deplores the recent rising and desires to make it plain that it has no connection whatever with it'.[101] Another Dublin based Anglican organisation, the Association of the Relief of Distressed Protestants, which had refused to take sides during the Dublin lock-out of 1913, now 'had no such inhibitions about taking sides', and denounced the rebellion.[102] The Presbyterians did not merely condemn it, but drew the lesson that it was the 'supineness of those responsible for the Government of Ireland which led up to this rebellion', though they did praise the 'energetic action on the part of the British Cabinet' in dealing with the rising, despite the 'calamitous expense both in lives and property'.[103] That property included Lower

Abbey Street Presbyterian Church, which with its church hall and other buildings was completely destroyed.

Bernard, the Bishop of Ossory, had become the Archbishop of Dublin in 1915, returning to the city where he had spent most of his working life as a Fellow of Trinity and Dean of St Patrick's. The war had already left its mark on him: 'I go to this new work', he wrote on the eve of his translation to Dublin, 'with a heavy heart, for my hopes are buried in Gallipoli. My dear son was killed on April 26th, the day after the Dublin Fusiliers forced a landing, leading his men in a bayonet charge.'[104] That the 1916 rising should take place in Easter week, which was also anniversary of his son's death, perhaps explains the acerbity of the letter he wrote to *The Times* early in May: 'This is not the time for amnesties and pardons; it is the time for punishment swift and stern.'[105] If Bernard had personal reasons to feel deeply the contrast between his son's sacrifice and that of the rebels, so too – within a few months – had many Ulster Protestants. 1916 was[106]

> tragic for Ireland because of the Easter Week Rebellion and the consequences of that wild adventure, and tragic for Ulster because of the terrible sacrifices made by her immortal Division at the Battle of the Somme on July 1. But in the latter case, where the fight was for the freedom of the world, and not in support of any selfish interest, political or ecclesiastical, the spirit of a brave people had never shone to better advantage then in this staggering, overwhelming calamity; and, while mourning for the noble dead, the whole Province rejoiced in the unsurpassed gallantry of its sons and the glory which their great deeds of valour had reflected upon an Imperial race.

Those words, written ten years after the battle in a history of the Belfast City Mission, illustrate how, after the war, the Somme and the Easter Rising 'had become the twin symbols of loyalism and treachery'.[107] The battle of the Somme opened on 1 July, the original date of the battle of the Boyne. Two thousand Ulstermen were killed; over 3,000 more had been injured. 'As once in Egypt, so then in Ulster there was scarcely a home where there was not one dead', wrote one Presbyterian clergyman evoking again the Old Testament imagery which had seemed so appropriate a few years earlier during the Ulster crisis.[108] The Mabin family from the Shankill Road lost two sons in the battle; every year since then, they have placed a memorial notice in the local paper, and have flown the Union Jack on 1 July. Others have done the same.[109] The Somme, and '[T]he blood spent at Thiepval, Passchendaele and in the bloody retreat of 1918 had given Ulstermen a new martyrology.'[110] In the struggle over the future of Ireland, each side – Catholic and Protestant, Republican and Unionist – would be haunted by memories of the 'terrible beauty' born in 1916.

On June 1918 the Presbyterian General Assembly was looking forward in hope to a time when 'by a complete victory the foundation be laid for a firm and lasting peace'.[111] That was not the outcome in Ireland. The government's attempt to ensure a post-war settlement focused from July 1917 on the Irish Convention, meeting in Dublin under the chairmanship of Sir Horace Plunkett. While there was some agreement on the need for a measure of devolved government, Sinn Fein had boycotted the proceedings, and the Ulster Unionists insisted on the exclusion of the north east from the jurisdiction of any such government. Archbishop Bernard sided with the southern Unionist leaders Midleton and Desart in contemplating 'a scheme of self-government for all Ireland', rejecting partition or the exclusion of Ulster.[112] But Archbishop Crozier of Armagh took a different stand, and his 'vote was usually on the side of the Ulster Unionists' negative response to any proposal for Irish self-government', even though he 'regarded the prospect of partition with horror'.[113] It was a split within the church that was mirrored in miniature within the Irish Guild of the Church. Its Executive Committee in June 1916 had condemned the rising; its Annual General Meeting in May 1918 overturned that resolution. This led to the President (the Bishop of Killaloe) resigning, and walking out in company of many members, while the rest settled down to hear 'an interesting paper on the Galatian Celts'.[114] By the end of the year a new breakaway organisation, calling itself the Irish Guild of Witness, had been formed to carry on what it saw as the traditional values of the organisation. The Irish Guild of the Church continued on its own way – a way well illustrated by its resolution in 1920 condemning the church for being 'so constantly identified with the reactionary forces in this country'.[115] Clearly there were deep tensions and divisions now within the church of Ireland itself. Naturally such tensions were also evident in the Convention. They were exposed even before it began, with Bernard suggesting that it open with prayer, even if the prayer were to be spoken by a Roman Catholic prelate.[116] Crozier and the Ulster members could not agree as that would be 'an admission that an Irish parliament (if there is to be one!) should henceforth be so opened'. So the Convention opened with out any prayers; and it ended in failure. When the general election of 1918 was called just after the armistice, the battle lines in Ireland were clearly drawn. At one extreme were the Ulster

Unionists, refusing to be ruled from Dublin. At the other, Sinn Fein, who annihilated the old home rule party, refused to attend Westminster, and set up their own Provisional government in Dublin. Years of violence were to follow before the churches could feel that normal times had returned. For the Irish churches, unlike the churches in Britain, the war did not end until 1923.

Partition, civil war and sectarian violence

In February 1920 the Government of Ireland Bill was given its first reading in the House of Commons, and in March the Ulster Unionists accepted home rule as inevitable, but only on the grounds that if Dublin got it, Belfast would get it too. 'It seems as if the Government would yield on the exclusion of Ulster and then what will become of us all in the South', the southern Unionist Lord Oranmore had lamented in his journal in 1914.[117] Now six years later the situation was even worse, for while Ulster was prepared to accept home rule, Dublin was not, and the war now being waged – mainly in the south – was an attempt by Sinn Fein to achieve what the men of 1916 had died for, an independent Irish republic. The war lasted until the end of 1921; the Anglo-Irish Treaty was a compromise whereby Northern Ireland would continue to have its home rule parliament in Belfast, while the south would get not independence but 'dominion status'. It betokened 'the wreckage of all our work together for Ireland', Lord Midleton lamented to Bernard, now Provost of Trinity.[118] But at least the compromise was accepted by the majority. A minority – led by Eamon de Valera – fought on until April 1923 in the hope of achieving an all-Ireland republic. Thus for four and a half years after the guns fell silent on the western front, they continued to blaze in Ireland, with consequences that were to last for the rest of the century.

Irish Protestantism faced two problems of violence in these years: in the south, civil war and reprisals; in the north, the growth of sectarian violence. If Protestants had feared that home rule would be Rome rule in the years before the war, when complete separation from Britain had been the dream of only a few republicans, they were now confronted by a far more serious situation. In the Catholic press, a 'consciousness that Ireland was a "Catholic nation" grew'.[119] This must be seen in the context of a resurgent Catholicism on the continent, where new Catholic states like Lithuania and Poland had emerged from the chaos of war and the collapse of the Russian Empire. Roman Catholicism was an ideological bulwark against the new threat of Soviet Communism.

Even in France the church regained a hold after years of anti-clericalism.[120] In vain did Irish Protestants latch onto the British government's exposure of the Pope's peace proposals of August 1917, which 'confirmed the opinion already held in all Allied countries that the chief object of the peace proposals had been to bring about a peace favourable to Germany'.[121] In England the Catholic Church was more self-confident now, attracting notable converts in the 1920s.[122] English anti-Catholicism was not dead, but it could no longer be so moved by the cries of Irish Protestants. The British were, it seemed 'sick and tired of Ireland', as Lord Riddell commented. There was 'apathy' and an 'unwillingness to face unpleasant facts'.[123] The threat to Protestantism – even what Mary Kenny has called the 'ethnic cleansing' which southern Protestants endured in these years – no longer evoked the moral outrage which accompanied less fearful incidents during the pre-war years. Then, Irish clergy had successfully spread the message in Britain to receptive audiences, who now no longer responded, even though Protestants felt more at risk than ever before.

At Fermoy, Co. Cork in September 1919, British soldiers formed up on church parade. Outside the Methodist Church, one was shot dead and three were wounded; the corporal seeking a doctor had doors slammed in his face. Indignantly – and fearfully – the Methodist authorities protested.[124] In July 1920 a Banbridge man with a distinguished war record, Colonel Smyth, a divisional police commander in Munster, was shot in Cork by the IRA, and local railwaymen refused to work the train that was to take his body home to the north for burial. But if these men had been legitimate targets in the IRA's war against the Crown Forces, other incidents fell into a different category. In May 1921 the 'Rosslea burnings', as they came to be known, were reprisals by the IRA in the little border town where 14 houses were burned down and two Special constables murdered. At the meeting of the Royal Black Institution of Monaghan and Cavan, a month later, the Revd R.J. Little claimed that this was another example of 'an alarming attempt to rout the Protestants out of the country'.[125] John Finlay, the Dean of Leighlin, was murdered in 1922. In April 1922 a series of murders in West Cork (the 'Dunmanway massacre') left more than a dozen Protestants dead, and the Revd Ralph Harbord, the local curate, badly wounded. Attacks such as these, Mary Kenny claims, 'have been cloaked, omitted from the official history books and even denied by the Protestant community themselves, who, possibly for their own survival, kept their own *omerta*'.[126] There seemed to be no safe havens for Protestants when the Unionist MP, W.J. Twaddell, was murdered in a Belfast street in April

1922, and a month later Field Marshal Sir Henry Wilson was assassinated on his own London doorstep. The Presbyterians lamented that 'the trend of events in Southern Ireland is leading to the gradual withdrawal of many Protestant families in that area'.[127] In one Presbytery membership had been reduced by 45% in seven years, and the other southern Presbyteries suffered almost as much. At the General Synod in May 1922 the Archbishop of Armagh lamented 'the reign of terror – I may say a veritable nightmare of violence and bloodshed – which has been in existence for some time in parts of our country, both north and south'. He went on to express 'the deep sorrow with which we heard of the shocking deaths recently inflicted on some members of our Church, and of the expulsions of many others. We do not know what may be behind these appalling outrages. But it is a sinister fact that, in one district, there should be so many terrible crimes of which the victims were all Protestants.'[128] Some 200 'Big Houses' were burnt down in the years 1920–3. The 'Ascendancy exodus' continued; the 'ruined homes – once centres of culture, patriotism and religion – stand in their desolation as grim reminders of what Irishmen have suffered at Irish hands in recent days',[129] depriving the Church of Ireland of crucial support especially in the deep south and west. Yet in the midst of it all, there were more hopeful moments. In June 1921 a new Arklow lifeboat was launched: Lady Alice Howard christening it with a bottle of champagne, the Roman Catholic priest blessing it, and the Church of Ireland rector dedicating it. 'The Protestants sang "For those in peril on the sea"; the Catholics, "Hail Queen of Heaven the Ocean Star".'[130] In some ways, even for the beleaguered aristocracy of the south and west, life went on as normal. In other ways, it would never be the same again.

Nor would it be in Northern Ireland, that fledgling country carved out of six of the nine counties of historic Ulster, its new devolved Parliament opened by the King in June 1921. Just as the new Dominion in the south would be launched six months later in the aftermath of civil strife and to a stormy passage through more than a year of civil war, so Northern Ireland saw the light of day amid violence. 'Protestants adopted a siege mentality, characterised by the belief that the Catholic threat from the south was compounded by the Catholic enemy within. This was symbolised in the slogan "For God and Ulster", a principle that the Protestant faith and the state were indivisible.'[131] Northern Ireland was indeed under siege: the southern government launched an economic boycott of the north in August 1920 which lasted until early 1922. Bouts of sectarian violence accompanied the birth of Northern Ireland, just as 'ethnic cleansing' disfigured the emerging dominion in the south.

Sectarian rioting shook Derry in the summer of 1920, and worse followed in Belfast where hundreds of Catholics were forced from their homes. Further violence followed in 1922 and 1923 – 'pogroms' inflicted on the Catholics in the north matching the 'ethnic cleansing' of Protestants in the south. Between December 1921 and May 1922, 89 Protestants were killed and 180 were wounded; 147 Roman Catholics died, and 166 wounded. There was no pogrom, said the northern Prime Minister, Sir James Craig: 'I repudiate that with all the language at my command.'[132] But it was certainly sectarian violence, and there seemed little recognition within the churches that this was not merely a political problem, but a social and moral one. Surveying the scene in 1923, the Presbyterians identified the social evils 'still with us' as 'Intemperance, Gambling, and Social Immorality'.[133] The Protestant churches would have to face all these problems in the years ahead, but it would be many years before they identified sectarianism as an equally corrosive evil.

The churches now operated not within one jurisdiction, but in two. 'Whatever the political future of Ireland', the Presbyterians declared, 'there is one point in which we must take our stand as Presbyterians – there must be no partition in our Church ... We are all members of one Church – the Presbyterian Church in Ireland.'[134] On that point, all the Protestant Churches were agreed. The border might now be a constitutional fact; it might even be drawn, in J.C. Beckett's phrase, 'in the minds of men', but there was to be no partition on the ecclesiastical map of Ireland. Indeed the borders between the churches themselves were now the subject of renewed scrutiny. The first formal meeting of the United Council of Churches in Ireland took place in January 1923. Just as there were no borders separating the churches in the north from those in the south, so perhaps the borders between the churches could be dismantled. That was the theory. It remained to be seen whether it could work in practice.

3
The Gospel and Society, 1900–1914

> The many-sided problems of social unrest come home to men's business and bosoms, and, therefore, touch every human interest. Not only the man in the market-place and the women in the home, but the economist and statesman, the philanthropist and moralist, find themselves compelled to answer its questions. The Christian Church, with its mission of healing the moral diseases and remedying the moral wrongs of the time, cannot escape the duty of facing the issues of our social ferment. To know the obligations of the Church and to be assured of its methods of service is an imperative duty.
>
> *W.M. Clow, 1913*[1]

The ills of society

'There is no more hopeful sign in the Christian Church of to-day than the increased attention which is being given to the poor and outcast classes of society.'[2] This was the verdict of the Revd Alfred Mearns, an English Congregational minister who in 1883 highlighted a new wave of social concern in the churches. Enhanced awareness of the needs of those 'in heathen lands afar' was matched by an increased concern for those in need in what William Booth called *Darkest England*.[3] Indeed Booth's own Salvation Army was an attempt to reach out to the 'unchurched masses' not only by preaching the Gospel, but by practical help. By the end of the century social problems had assumed a central position in Christian thinking, and men such as Westcott, Figgis, Scott Holland, Keeble and Clifford were giving them a prominence in their preaching and writing that actually shifted the centre of gravity of Protestant thinking. It has been claimed that such a shift was inspired – at least initially – more by concern for falling church attendance than by a genuine awareness of vast new social problems.[4] Certainly the development of Irish Protestant social thinking seems to bear this out, and to have followed the pattern of development of the churches in Britain, though with a time lag.

The industrial revolution had created a new social order. Owen Chadwick has noted that '[S]o far as the churches or chapels possessed the allegiance of the working class of England and Wales, they lost that allegiance when the country labourer became a town labourer.'[5] Ireland however remained a largely rural society outside the north-east. It was industrial Belfast that would focus the problem of the 'unchurched masses'. The Bishop of Connor spoke in 1910 of 'the great burden laid upon us by the overwhelming influx into our city of the labouring and artizan classes ... They have come from distant homes where their clergy knew and sympathised with them, and they find themselves, through no fault of our city clergy, but through their poverty, almost forgotten in this great lonely city of ours – little wonder that they gradually in large numbers lose their church connection.'[6] By 1910, all the churches had begun to recognise that they faced challenges which could satisfactorily be met only by employing new methods and by presenting the Gospel with a different focus and different emphasis.

The traditional approach had been expounded by the Presbyterian minister, Henry Magee, head of the Mission in Dublin, when in 1885 he identified the object of his mission's work. It was 'to draw the heathen from sin to holiness, and from love and service of the world to love and service of God'. Presumably all Christian teaching in every age has this object, but the methods he identified indicate a state of thinking which predates the age of the 'Social Gospel': Christian co-operation, 'offering Christ', colportage, the use of the press, and preaching in Irish.[7] The work of the church was seen in terms of its mission to the individual, and social problems became relevant in so far as they prevented the individual from responding to the Gospel. In the second half of the nineteenth century 'drink' was seen as the problem *par excellence*. 'Intemperance continues to impede the progress of the Gospel', as one of the Belfast Town Mission workers reported in 1867.[8] Throughout the 1870s, 1880s and 1890s, this was a theme in pulpit and on platform. The temperance movement in Ireland was not a new phenomenon: the

Catholic Father Mathew and the Presbyterian John Edgar were great temperance advocates in the earlier part of the century. The Irish Temperance League – based in Belfast – had been founded in 1854, and by the end of the century employed a full-time secretary and assistant secretary, and involved itself in all aspects of the anti-drink campaign. Although non-political and self-proclaimedly non-sectarian, it was in effect a pan-Protestant organisation. It worked alongside the temperance committees of the various churches, as well as in co-operation with the numerous other temperance movements in Ireland – the Army Temperance Society, the Bible Temperance Association, the Order of Good Templars, the Band of Hope, the Irish Sunday Closing Association and the Rechabites among others. The churches were happy to support an organisation which – via its weekly committee meetings – could concentrate on combating the demon drink. It helped the churches themselves set up temperance missions;[9] it employed lecturers to spell out to children the dangers of drink;[10] it ran coffee stalls and temperance cafes in Belfast, and undertook the catering at Harland and Wolff's and at the students' union of the Queen's College.[11] The churches had neither the time nor the resources to fight drink on such a broad front, except perhaps via their city missions. The Methodist one, an enthusiastic supporter claimed, had 'done more to ruin the drink traffic in Belfast than perhaps any other agency'.[12] The churches certainly played their part, and the Methodists were most prominent, campaigning not merely for temperance, but for total abstinence. 'We counsel you', Conference proclaimed in one of its many statements on the subject, 'to refrain entirely from the use of these drinks, except in the case of extreme necessity.'[13] The Presbyterian Church advised its members to set 'an example of personal abstinence and of entire separation from the drink traffic and from the drinking customs of society'.[14] The reason was not a killjoy one, for as the General Assembly in 1891 proclaimed, almost as a self-evident truth, 'intemperance is at the root of most of our poverty, lunacy and crime'.[15] This was an interpretation – a sociological observation – which had become widespread in Britain in the middle years of the nineteenth century. Drinking was the problem; it was not a symptom, it was the disease. And, as Peter Mathias has said, 'the vocabulary was that of sin and temptation; there was no deep-seated institutional evil to be eradicated'.[16] The success of the temperance movement, it has been claimed, was greater in Ireland than in Britain or in the United States. More people joined in. Social attitudes, right up to the end of the twentieth century, reflected a higher percentage of total abstainers than in any other western country.[17]

One related issue began to loom large within the churches in the latter part of the nineteenth century – the question of the use of wine at holy communion. In 1875 an Irish Sacramental Wine Association was founded by the Congregational cleric, John Pyper, devoted to 'the promotion of the Divine glory, through the removal of the intoxicating cup from the Table of the Lord and the entire separation of the Church from the liquor traffic'.[18] After little initial impact, the issue became a major bone of contention. The Anglican *Irish Ecclesiastical Gazette* took up a predictably traditional stance, declaring that the use of non-fermented wine would 'destroy the truth of the sacrament, which might as well be administered in milk or water as in a non-fermented ingredient propounded probably in a chemist's or an apothecary's establishment'.[19] The Presbyterian General Assembly refused to sanction the use of 'unfermented juice of the grape' in 1875, but by 1898 it had been declared optional.[20] That did not lay the problem to rest however. First Carrick Church session was divided over the issue for several years, until eventually in 1893 the Moderator of the Synod 'agreed that he should ask the Congregation as a special favour, in order to preserve unanimity' to accept the use of unfermented wine.[21] At St Enoch's, the great 2,000 seater Presbyterian church in Belfast, the controversy was more bitter. The minister, Charles Davey, who himself favoured unfermented wine, tried to keep both sides happy by introducing two tables at communion services, leading some members of his congregation to complain to the General Assembly. It found for the minister, but the strain was so great that 'in 1897 he was compelled to take a special holiday in order to regain his health'.[22] More surprisingly, the Methodists were also split on the issue, the 1875 Conference deprecating 'any attempt to disturb the minds of the people' by raising the question. But in 1878 use of non-alcoholic 'wine' was made optional, and by the turn of the century, was universal.[23] This follows closely the chronology in England, where by the late 1890s a temperance historian could claim that 'probably a majority of Nonconformist churches now use a non-alcoholic wine'.[24] It had been a divisive issue, from which the Irish Temperance League tried in vain to stand aloof.[25] It also alienated many Anglicans from the broader temperance crusade. Curiously its influence was felt later in the twentieth century, when it became one of the emotive obstacles to Anglican–Methodist Union in England in 1967.[26]

In a wider perspective, however, the importance of the 'drink question' for the churches lay in its provision of a focus for Protestant social analysis in the latter part of the nineteenth century. Drink was seen to be at the root of almost every social evil. A shift in emphasis is however noticeable in the 1890s. The Irish Temperance League itself began highlighting not merely the personal reasons why a man might drink, but the 'social conditions, institutions, laws under which he is placed', admitting that 'the causes of his vice are not wholly his but are also partly ours'.[27] The General Assembly took up a similar theme when in 1904 it noted the 'untouched and immoral social conditions in which many of the very poor live', and urged 'on municipalities and Urban and District Councils the better housing of the poor'.[28] The emphasis on individual sin, and the moralism which attributed drunkenness to the waywardness of the individual, and the poor social conditions in which such a person lived to the evil of drink which pulled him down, was giving way to a different kind of analysis. This shift was part of a larger movement in thought, which had already taken place in England.

This shift in focus may be seen in miniature in Moody and Sankey's Irish crusades. The reaction of the churches to the 1874 mission was enthusiastic but highly 'spiritual'; saving of souls was the emphasis. A curate reported at a prayer meeting in 1883 that 'it was as a result of Moody's Dublin Mission in 1874 that ten men including himself had gone forward for ordination'.[29] In the 1892 Moody mission, however, 'a determined attempt was made ... to reach those who never went to church'.[30] This enhanced sense of outreach helped to draw the attention of churchmen to the new urban challenge which had long been felt in England. The last years of the nineteenth century were full of agonising over 'the extreme difficulty of reaching the masses'.[31] The generally agreed figure was that some 40,000 out of 200,000 people in Belfast never went to church.[32] In addition, the church-going population was itself increasing, the Presbyterians computing that 'we would require a new church erected each year to meet the increase'.[33] The result of the awareness of a need, not only to keep pace, but also to reach out, was that the late nineteenth century was the greatest period of church building in Belfast's history. The Methodists opened their Belfast Central Mission in 1894; the Presbyterians founded their Shankill Road Mission in 1898. TCD graduates were fired with enthusiasm in 1888 for founding an urban mission, following the example of the English public schools and Oxbridge colleges, though it did not materialise until 1912.

It was not long before the challenge of numbers turned into something else. It was picturesquely put by the country clergyman who declared that 'Christianity (so-called) that is not practical in everyday life is really nothing but sing-song twaddle'.[34] The churches had long made pronouncements about issues which affected everyday life – notably in the fields of education and temperance reform. But from around 1890 onwards, their area of interest widened into a profusion of official statements on Sabbath observance, smoking, gambling and sexual vice. These focus on the sins of the individual. Structural analysis of the causes of poverty, crime and vice was slower in coming. Roman Catholics also held to the view 'that social reform is to be accomplished more by a reform of individual characters than that of social structures'.[35] However a deeper analysis became evident during the reign of Pope Leo XIII, whose corpus of writings, and especially his 1891 encyclical *Rerum novarum*, gave his church a lead in facing up to the problems of modern society in terms of workers' rights, trade unions, and the need for a minimum wage. Irish Protestants had no such external source of authority – though having one did not necessarily mean that it would be heeded; the Irish Catholic Church was slow to pick up on the new papal departure. Protestants could look to the example of churches in England who had battled for longer with the growth of an urban, secular society. To such a church the Revd William Nicholas, a prominent Irish Methodist, delivered the Fernley Lecture, when he spoke at the British Methodist Conference in Cardiff in 1893. His subject was 'Christianity and Socialism', and his lecture breathed the confidence of the mid-Victorian liberal, who believed that 'we are advancing, perhaps not very rapidly, but yet most certainly, to the day of perfect individual, civil and religious liberty'.[36] As to social evils, society was not to blame. His explanation was a traditional one; it was one that he had grown up with. He had been ordained in 1861; his views reflect the climate of the age. In religious terms it had been the aftermath of the great 1859 Revival which swept Ireland, and politically it was the hey-day of mid-Victorian *laissez-faire* individualism. His remedy too was a traditional one: that drunkenness, idleness and dishonesty (the major social ills as he saw them) would cease when the principles of Christianity were accepted 'in the hearts and lives of men'.[37] The Irish Methodist newspaper hailed his lecture as a triumph,[38] but the *Methodist Times* in England branded it as an example of extreme individualism.[39] For English Methodism had moved beyond that point. Nicholas' lecture is an epitaph to an era of individualistic social analysis. At the turn of the century, his book was, for Irish Methodist ordinands, the only required reading which had any bearing on contemporary social questions; by 1913 it had been joined in their reading lists by F.G. Peabody's *Jesus Christ and the Social Question*. Published by a Harvard professor in 1901, this book was one of a flood of such works that around the turn of the century in Britain and America were trying to relate the teachings of the church to secular movements for the improvement of social conditions.

By the turn of the century, Irish attitudes were being influenced by this changing intellectual climate. Edward Norman notes that 'collectivist attitudes within the leadership of the Church had become so effective that by 1897 ... the Lambeth Conference of that year appeared to represent almost no other view'.[40] Although Lambeth Conferences had no authority within the Church of Ireland (or any church of the Anglican Communion) they invariably made an impression on the bishops who attended them. In 1898 the Down Diocesan Synod, in response to the recommendations of Lambeth, discussed the setting up of a social service committee, which was finally formed in 1902. In 1900 the Dublin Synod formed a committee 'to deal with social problems from a Christian point of view'. In 1903 the Presbyterian General Assembly set up a committee 'on the work of the church in industrial centres', and the Belfast Presbytery started it social service committee in 1907. Behind this organisational advance, new thinking was in the air. In 1898 the Revd M.F. Bovenizer caught the mood when he told his fellow clergy that 'the child needs, not the Christianity which prepares for death, but the Christ who prepared for a life equipped with pure and noble energies and ideals, in order through the power of such a Christ possessed life to accomplish a chapter of manly and selfless service for men'.[41] Two themes here are of interest. One is the emphasis not so much on traditional doctrines, as on a commitment to the personality of Christ, a distinct trend in contemporary English nonconformity. The other had similar contemporary relevance – the idea of 'service'. This was translated into action in the social service committees of the churches. And both themes are present in the Irish Conference's pastoral letter which proclaimed that innumerable schemes 'for the amelioration of social conditions are on foot; you may call them socialistic if you will, but whether such movements are going to be Christian or anti-Christian rests with the Church of Christ'.[42] The Dublin Presbytery outdid the Methodists by emphasising 'the duty of the Church of Christ to assist in liberating those who suffer from the heavy burden', in this case of long working hours.[43] This first faltering step towards a liberation theology, as it would later be called, was a significant development. The General Assembly's Committee on the work of the church in Industrial Centres produced its first report in 1904, and carried forward the idea that both theoretical and structural changes were needed.[44] It identified social and industrial developments to which the church had not adequately responded, and which had weakened the family unit, with the result that men fell more easily into habits of sloth, drinking and gambling. Here was an analysis that was not contingent upon individual sin; here too were practical solutions. It spoke of the provision of clubs, libraries and evening classes. It would increasingly put pressure on government, the only agency capable of coping with social problems too large for the efforts of individual churches, let alone individuals within the churches.

The Church of Ireland was active too, and did not confine its new-found interest to a 'knowledge of the social evils that surround us, and of the causes that promote or aggravate them'.[45] The Guild of Alexandra College concerned itself with housing, while the Church of Ireland Labour Home and Yard, founded in 1899 and managed by the Church Army, operated in the working-class Ringsend area of Dublin. In 1902 the Belfast Women's Workers Settlement was founded to 'provide a residence and centre for ladies who are desirous of helping in the work of the Church of Ireland in the City of Belfast', and in 1912 – at last – a Trinity College Mission was opened in the city. *Laissez-faire* was abandoned. 'Let us then as Christians demand from the legislature a strong policy of social reform', the Archdeacon of Down proclaimed; 'Let us insist on municipalities planning away the slum districts, and preventing the awful death rate among the children of the poor.'[46] Many social issues, especially in the period 1910–14, attracted the attention and campaigning zeal of the churches – unemployment, labour exchanges, national insurance, housing, provisions for legal aid, better conditions for children, changes in the poor law system, and greater controls over the 'white slave traffic'.[47] The demand for a living wage for workers is particularly interesting, because it was not a call made from a paternalistic or charitable standpoint, but rather a demand for social justice – for a fair day's wage for a fair day's work.[48] The individual's spiritual condition was not an issue. The churches had moved a long way from the individualistic, highly spiritualised social comments of a few decades previously.

This stand on a living wage did not, however, bring the Protestant churches into any closer contact with the labour or socialist movement in Ireland. In England, the links between nonconformity and labour were strong and deep, and the tradition of Christian Socialism had a place within the Church of England. Most of the prominent trade unionists and labour politicians of the late nineteenth century owed much to their religious upbringing, and brought religious insights to bear upon their political activities. But Irish Protestants had, since the 1870s, looked askance at the 'socialistic' tendencies of the Land League, and at the 'congeries of men of all creeds, any creed and no creed, secularists, socialists et hoc genus omne', who made up the home rule movement.[49] The peculiar political context in which Protestantism operated in Ireland in the half century before World War I – with opposition to home rule as the cardinal tenet in Protestant thinking – militated against a significant alliance between the churches and any political or social movement which was not devoted to that end. So too did the conservatism of the Belfast working class, expressed in its membership of the Orange Order. It was at an Orange service that one cleric attacked the Liberal government in words obviously calculated to appeal to his hearers, when he criticised its 'rushing the nation to financial bankruptcy by its hasty and unsuited legislation and its pandering to Socialistic demands'.[50] Despite the brief appearance of the Independent Orange Order in the years 1902–7, when it had seemed possible that a working class, anti-employer movement might develop in the north east, traditional loyalties and fear of home rule soon regained the upper hand.[51] The Belfast Trades Council, a non-sectarian trade union and labour body, did not develop the sort of links which similar bodies in England and Scotland developed with Church or Chapel. Indeed in 1905 it denounced any attempt to forge such an alliance between the 'free churches' and the labour movement, and in general its minute books carry note of bitterness with regard to the churches.[52] The feeling was that the churches were all too culpable in blunting working-class unity by fostering sectarianism.[53] There was an element of truth in this. When one Presbyterian minister tried to bring Christian influence to bear during the 1907 dockers' strike, he found his interference resented, most especially by the management, whom he thought were all too willing to stir up religious strife in order to counter radical demands.[54] Of course, the fact that the labour movement in Ireland was headed by James Connolly and James Larkin, both Catholics and both opponents of British rule, made it unlikely that any coherent links would develop between it and the Protestant churches.

Church-related organisations

The influence and activity of the Protestant churches in Ireland cannot be measured simply in terms of their involvement in the political process. Nor is it possible to understand their influence simply by reporting their official pronouncements. Churches were centres of local activity, places where people met, and not only worshipped together but also undertook other activities together. Sunday schools were important in this respect. Throughout late Victorian times, and for perhaps the first half the twentieth century, the Sunday school was a feature of the life of very large numbers of children throughout the United Kingdom. Even if parents were not themselves church-goers, the children were sent to Sunday school. Maybe, of course, as one boy remembered of his experience in Belfast in the 1920s, it was 'as much to get me out of the way as to have me save my immortal soul'.[55] There was also the suggestion that Sunday school 'provided little more than an opportunity for working-class adults to enjoy their conjugal rights in peace'.[56] But there was more to it than that. In 1893, just over 93,000 pupils were enrolled in Presbyterian 'Sabbath' schools, almost a third of those being in Belfast. They were taught by almost 10,000 teachers, giving Sunday schools on average a better pupil:teacher ratio than most primary and secondary schools. At St Enoch's, Belfast's largest Presbyterian Church, there was a Sabbath school of 2,300 pupils and 120 teachers. That so many adults, week by week, there and elsewhere, gave time to presenting Christian truths to their pupils was significant, and not only in religious terms. Children (and young people) learned to read aloud, to speak in public, to sing together, and to behave in a disciplined way. The social impact of all that should not be underestimated. In one Presbyterian church – probably in many – it was via the Sunday school that the practice of singing hymns gradually caught on, to be eventually introduced into church worship as well.[57] In one Congregational Church Sunday school in Belfast, which started in 1894, numbers rose from 19 to 57, with three teachers, within the year. The spirit of its organisers is well captured in their first annual report: 'We would offer our sincere thanks and praise to Him who so marvellously opened up the way for us to labour in this corner of His vineyard and pray that we may soon see the fruit of our work.'[58] Within two years, the school had become the base for a Band of Hope, a night school and a sewing class. Poignantly, its records contain the information that on Christmas Day, 1895, Jim Whyte died. A member of the Sunday school, aged just seven, he was buried the next day, and probably his only memorial is contained in the Sunday school records: 'he was a bright intelligent and quiet lad and was well liked by all'. Many boys and girls, particularly from the poorer sections of society, found in Sunday school a haven where adults took an interest in them, organised activities specifically for them, and tried to speak their language. The Crusaders was another movement geared to this age group, founded in London in 1906. It reached Dublin in 1914, though the first group did not start in Belfast until 1927. The aim of giving each member 'a Book in his hand, a Saviour in his heart and a purpose in his life' sums up the philosophy of many committed adults whose evangelism and enthusiasm took the message of Christianity out of the pulpit and into the lives and experience of countless young people.[59]

Sunday schools did not cater only for young children. One of the largest, Rosemary Street, Belfast, which still had 2,000 pupils on the eve of World War II, had begun around the turn of the century with a collection of dock workers – whom their teacher called the 'forty thieves'.[60] There were other schools where the average age remained around 21 until the 1920s.[61] More commonly, older 'pupils' enrolled in a Bible class, such as the one at Christ Church, Lisburn, which was held to be the largest in the United Kingdom, and attracted an average attendance of 500 at the turn of the century.[62] For many children and young people, Sunday school or Bible class also provided the opportunity for holiday activity on a modest scale. Particularly for city children, the Sunday school picnic or excursion might be the only time in the year that they escaped from their mean streets. It was not always a joyous occasion: in 1889 the Armagh Methodist Sunday school outing of 1,000 people ended in disaster when the train was derailed en route to Warrenpoint, killing 80 of the passengers and injuring 200. One small comfort after the tragedy was the sponsoring of an aid fund not only by the Methodist Church, but also by Archbishop Knox and Cardinal Logue. Sadly, there had been a Sunday school incident two years earlier that was divisive rather than unifying in its effect. A party of 80 boys and girls, and 19 adults had set out from Westport for a picnic; on their way home, they were attacked by some 150 people 'who proceeded to groan and hiss, and throw volleys of stones and turf'. Injuries rather than deaths resulted, but the incident became part of the Protestant armoury in the anti-home rule campaign.[63] A similar incident, as we have seen, added to the highly-charged atmosphere of Ulster in the summer of 1912: the 'Castledawson outrage'. The minister concerned appeared as a witness at the trial of the men responsible, and pleaded for mercy.[64] That was a rather better lesson for the Sunday school children than the sectarian animosities which were stoked up by press and pulpit during that eventful summer.

The Christian Endeavour movement provided yet another opportunity for young people, and young adults, to learn about the Christian faith and to meet socially. Founded in Maine in the USA in 1881, its first Irish branch began in Agnes St Presbyterian Church in 1889, the 39th society registered in the British Isles at that time.[65] Others soon followed, and by the end of the century there were over 100 societies in Ireland, and by 1907 there were 10,000 members. It was an ecumenical movement among the young: its motto, 'For Christ and the

Church', did not refer to any particular church. Of its first 30 presidents in Ireland, twelve were Presbyterian, eleven Methodist, three Anglican, two from the Society of Friends, and one each a Baptist and a Congregationalist. The Presbyterians and Anglicans had societies organised into unions within their own denominations, whereas the Methodists and others affiliated their societies directly to the national CE Union, to which the denominational unions were also affiliated. Its first national convention in Ireland was held in 1894, and its newspaper – soon to be called *The Irish Endeavourer* – was founded a year later. Praising the movement in 1900, the *Irish Presbyterian* noted that already, after just 20 years, it could mobilise 3.5 million young people worldwide. For many in Ireland, it became the channel for their Christian enthusiasm, and also in frequent cases the stimulus for them to go on to full-time ministry in their respective churches. One Presbyterian minister traced the process whereby through the CE he had progressed in his Christian witness: first, by reading a passage of Scripture; then later by leading in prayer 'very haltingly'; next by reading a paper at one of the weekly meetings; then by speaking in the open air 'in front of the public house in the village'.[66] He went on to enter the ministry; in 1963 he was Moderator of the Presbyterian Church. For the Methodists too it 'proved a fruitful seed plot for the growth of candidates for the ministry'.[67]

There were, of course, youth organisations which were much more visible. Members of Christian Endeavour were not instantly recognisable, though they could sport a small lapel badge. Similarly, Sunday school children looked like any other children, as did the men and women who attended Bible classes. But the 'uniformed organisations' also played a role in evangelising and educating young people, and in addition gave them a much more public persona, a pride in their appearance, and indeed a significant role both in their churches and in society at large. And although there were organisations for both boys and girls, there is no doubt that boys were the prime target, and the movements launched on their behalf are the more significant aspect of this development in late Victorian and Edwardian Britain and Ireland. Adolescence, it has been claimed, was 'one of the social effects of the mid-Victorian "reform" of the public schools'.[68] By late Victorian times adolescence had become part of a much longer process of growing up. The parliamentary reforms of the Victorian era had outlawed child labour; compulsory education had become part of the experience of young people, who no longer faced wage earning at the age of ten or younger. For public school boys the transition from childhood to manhood was becoming standardised: preparatory school, public school, university. For young people – and especially boys – in the less wealthy sections of the community a similar transition phase was now in place. For many working-class and middle-class youths, the churches' 'uniformed organisations' provided the vehicle whereby boys could travel that road from childhood through adolescence to young manhood, and experience some of the camaraderie, discipline, adventure and training which public schools provided for the more privileged. The churches had always stood at the crucial moments of the life of families – in baptism, confirmation, marriage, burial. In late Victorian times, they began to do so in that more secular and more drawn out transition from boyhood to manhood.

One of the most significant youth organisations, and one of the earliest, was the Boys' Brigade. Founded in Scotland by William Smith in 1883, it had the overtly evangelical aim of spreading 'Christ's kingdom among boys', and 'the promotion of habits of Reverence, Self-respect and all that tends to a true Christian manliness'. The founder explained that 'by associating Christianity with all that was most noble and manly in a boy's sight, we would be going a long way to disabuse his mind of the idea that there is anything effeminate or weak about Christianity'.[69] In 1888 the first company was set up in Belfast, attached to St Mary Magdalene (Church of Ireland); the driving force behind it was the Sunday school superintendent, William McVicker, whose son was to become a leading personality in the world-wide movement. Within five years the BB in Ulster had spread to ten provincial towns, and McVicker had organised 18 companies in Belfast. Meanwhile the Revd William Dobson, of St Matthias, Dublin, started the first company in that city and by 1893 there were 21 companies, almost all attached to Anglican churches. Thirty-five years later there would be 71 Irish companies in all. Within and between these companies, there was much emphasis on competition, and around the turn of the century, 'Drill, gymnastics, cricket, dumb-bell, bugling, photographic, swimming, tug-o-war, and hyacinth growing competitions, added zest to the normal company programme.'[70] Other similar movements soon took root in Ireland: a specifically Church of England equivalent of the BB, the Church Lads' Brigade, was founded in 1891, while a nonconformist equivalent, the Boys' Life Brigade, started in Scotland in 1899. They all had in common an emphasis on 'bearing' and on smart appearance, though as the arrangements for the Belfast Church Lads' Brigade Challenge Shield made clear in 1908, the actual quality of the clothes was 'not to be taken into consideration'.[71] It was the development of an *esprit de corps* that was important; elements of uniform helped to enhance this. The BB pillbox hat and white belt soon became well known and survived almost unchanged until 1971.

This emphasis on bearing and appearance was not merely a concern about superficialities. These outward signs were held to reflect an inward attitude of discipline and pride, and were the marks of a vigorous young manhood. Significantly, the BB's aims were broadened ten years after its founding by the inclusion of the word 'discipline'.[72] There were strong currents of opinion around the beginning of the twentieth century concerned with the 'purity of the race', with 'national efficiency', and with Social Darwinism. As almost invariably happens, churchmen were affected by these contemporary ideas, and were involved in applying them – often with evangelical zeal – to the style and content of their gospel.[73] In Britain, Germany and the USA, eugenics societies sprang up, with a concern about the survival of the race. The eugenics movement in Ireland attracted people like W.B. Yeats, Lady Aberdeen, and members of the Guinness family, and the Belfast Eugenics Society, founded in 1911, was chaired by Bishop D'Arcy. The problem as regards the survival of the 'race' as a healthy and dynamic motor in national life was perceived largely as an urban one, and particularly focused on young men. 'Too much of young Belfast', the *Northern Whig* proclaimed in 1902, 'crouches and shambles rather than walks at the present moment; too much of young Belfast spends its evening's leisure on the streets ....'[74] In the same year, the Earl of Meath declared that there were lessons to be learned from the Boer War, that 'every healthy British lad should be so trained in his youth that, if the occasion ever arose, he would without delay take an active and efficient part in the defence of his country'.[75] Out of that same concern came the Boy Scout movement, founded in 1908 – with its emphasis on healthy outdoor exercise, self-reliance and patriotic fervour. While less overtly Christian than the BB, its concern that boys should do their duty 'to God and the King' commended it as a youth movement to the churches. The first Irish troop started in Belfast in 1908. Within a couple of years the BB 'had to face the problem of the attractiveness of the Boy Scouts' organisation', and indeed scouting was added to BB activities.[76] Whichever youth organisation a boy belonged to, he was exposed both to Christian teaching, and to physical training and healthy activity. The young Presbyterian minister, David Corkey, devoted much of his ministry in Shankill Road, from 1906 to 1911, to working with boys. Soon after he arrived, he had started two football teams and a gymnasium. By 1908 he had plunged into work for the BB. He recorded amusingly in his diary the success of a talk he gave his boys on 'Cleanliness is next to godliness': 'Went away for a fortnight's holiday, came back and every mother was complaining that the boys were running to the baths every moment they had. One complained, "James will soon have himself washed away".'[77] A true urban missionary, Corkey set up house with his sister in Percy Street, in working-class Belfast. There, his library was open to the boys; there were hill rambles, drill instruction and football. There was a Bible class, and a huge attendance of 'his boys' when he preached at the Albert Hall. He had, his sister writes, 'a vivid sense of the end he wished to achieve, which was nothing more than to recreate the sane, vigorous, healthy Christian manhood which he himself possessed'.[78] There must have been many young clergy like him in the pre-war towns of Ireland. How many of them perished, along with their 'boys' in that war? David Corkey himself survived it for just a few years, but his health was broken and he died at the age of 41.

Girls were catered for as well. A female equivalent of the BB – the Girls' Brigade – was started in Ireland in 1893. It grew out of a girls' singing practice – at Sandymount Presbyterian Church in Dublin – which was preceded by some warm-up drill. Some of the girls had brothers who were in the BB, and suggested that their group should be called the 'Girls' Brigade'.[79] It was an Anglican, the Revd. E.C.H. Crosby Lewis, who presided at a meeting in October 1908 which established the Girls' Brigade as a national organisation. In Ulster its companies were mainly attached to Presbyterian and Methodist churches, while in southern Ireland it was mainly Anglican. A female equivalent of the Boys' Life Brigade started up at Thomas Street Methodist Church in Portadown, and spread throughout the Methodist and Presbyterian Churches. A specifically Presbyterian organisation, the Girls' Auxiliary, was founded in 1911 with the aim of 'thought, prayer, comradeship and service' at its heart.[80] Girls' Friendly Societies, attached to Anglican churches since the first Irish branch was opened in Bray in 1872, provided a social focus for young women, without the uniformed aspects. The Presbyterians also tackled the problem of young women who worked in Belfast and lived away from home. Their Presbyterian Women's Union, founded in 1905, provided rooms in Church House where Bible classes were run, and evening meetings were organised for singing, sewing and other activities. Young women could go there to read magazines and papers, and to take afternoon tea. It would be true to say however that in the churches, as in the nation as a whole, there was not the same concentration on the needs of girls as of boys. While there were church organisations which catered for girls, and while particular efforts were made to deal with specifically female problems – for example rehabilitating young women after their release from prison, or dealing with unmarried mothers – it was males who were felt to present the greatest problem, and whose training and disciplining were most urgently required for the sake of the 'race' as a whole.

Every church building in Ireland was not merely a place of worship, but also in many ways a social centre – a focus for activities for people of all ages. When the Rector of St Barnabas, in Belfast, presented his report in 1912, he carefully recorded the various facets of the work of his church.[81] There was a Gleaners' Union, raising funds for the Church Missionary Society. There were Sunday schools as well as a day school. There was a Band of Hope, mobilising the young people in the temperance cause. For boys there was a Church Lads' Brigade, and for boys and girls a children's choir. A Bible class catered for the needs of young adults, and there was a Men's Institute. All age groups were involved in raising money for the Orphan Society. In an age before television and cinema, and in a city where public houses were hardly respectable, churches could provide activities for the whole family. When the new Methodist Belfast Central Mission opened in 1894, it had a large hall to seat 2,500 people and a minor hall seating some 500. In addition there was a public reading room and a gymnasium. There were working men's club rooms and a girls' 'parlour'. There was a 'City Arab Shelter' – a place of refuge for the orphans and waifs and strays of the city. There were committee rooms and class rooms. Even more than the normal church, its activities were more than just religious. Popular Saturday night entertainments were laid on: bands played, choirs and soloists sang, readings and addresses were given, plays were performed. Being a city 'mission', its layout and facilities were specifically geared to a ministry summed up in a 1902 leaflet:[82]

1. To rescue non-churchgoers.
2. To feed the poor and clothe the naked.
3. To rescue and uplift the down-trodden children of the slums.
4. To rescue from the haunts of vice and sin.
5. To rescue drunkards from the public house.
6. To help the working man when out of employment.
7. To make Belfast like unto the City of God.

There was no individualistic narrowness here. The 'social gospel' was taking root in Belfast.

More modest in its objects, but active in every Church of Ireland diocese by 1906, was the Mothers' Union. It was founded in the

Winchester diocese by Mary Sumner, whose husband was rector of Old Alresford. He ran his YMCA and men's Bible classes. His wife began to hold meetings for the women, which soon outgrew the rectory drawing room. 'What can be done to raise the national character?', she was asking at a Congress in Portsmouth in 1885 which marks the founding of the Mothers' Union. Just as the training of young men would improve the race, so too would the cause she espoused: 'Let us appeal to the mothers of England. It is the mothers who in great measure can work the reformation of the country.'[83] Just two years later, Anabella Hayes, wife of the rector of Raheny near Dublin, started the first branch for the mothers of Ireland. By 'something of a miracle' the idea spread from parish to parish,[84] with a Mothers' Union Council for Ireland created in 1903, and a presence established in every diocese by 1906. It was well patronised: the diocesan presidents included Lady Stronge in Armagh, the Countess of Belmore in Clogher, the Dowager Marchioness of Dufferin and Ava (who was also the National President) in Down, and a host of other titled ladies. The 'Establishment' ethos lingered. At least the Presbyterians could, from 1906, point to a Presbyterian Viceroy. Although he was a home ruler (which took some of the shine off this Presbyterian 'first' – and last, as it happens) he was devoted to good works, and particularly keen on the Boys' Brigade. His wife led a crusade against tuberculosis, which, as Hubert Butler commented, 'was one of those rare and blessed battle-cries, like co-operative creameries and village halls, which appeared to have no political or religious implications'.[85] In fact much – indeed most – of what was achieved by the churches in the early years of the century was entirely non-political. The worship in the churches might underline and reinforce the Protestantism of those who attended, and Sunday schools might help create in the minds of the young a clear appreciation of their 'Reformation heritage'. Yet in the formation of character, and in the multitude of activities geared to helping those in need both outside as well as inside the churches, the Protestant denominations played a vital part in the lives of many people. The principle of 'voluntarism' – the emphasis on individual charity and the reliance on initiatives from churches and societies – was no longer seen to be enough to cure the ills of society as the twentieth century dawned. But it had a part to play. The churches began to accept the need for greater state intervention; but they continued to believe in another dimension. In a defence of the Church of England in 1886 against the threat of disestablishment, the Earl of Selborne expressed it well: 'The Church is a Society placed by its Divine Founder <u>in</u> the world, though the spirit by which it is, or ought to be activated and animated is not of the world. From the world it neither is, nor by any possibility can be entirely separated.'[86] While Irish Protestantism is better known for its political stand than for its ethical and social teachings, the focus of its work – throughout the twentieth century – was not political at all. Its spiritual and moral guidance meant more to it, and to its members, and had more effect on their lives. But of course there is another church-related organisation which has meant that Irish Protestantism is widely seen in a political light. The Orange Order deserves a mention.

The Orange Order – a brief note

The Laws and Ordinances of the Orange Order spelled out the basis of the Institution.[87] It was 'composed of Protestants, united and resolved to the utmost of their power to support and defend the rightful Sovereign, the Protestant Religion, the Laws of the Realm, and the Succession to the Throne in the House of Brunswick, BEING PROTESTANT'. It had no formal institutional or constitutional links with the Protestant Churches, though for its first century up until disestablishment it had been strongly Anglican in membership. And while the Order had previously appeared to Presbyterians to be a means of propping up the Ascendancy, it was now seen as another plank in the anti-home rule platform. Doctrinally the beliefs of the Orange Order were impeccably pan-Protestant, and its devotion to the principles of the Reformation, to the Union, to law and order and to freedom of conscience, summed up the way Irish Protestants liked to see themselves. Their first united presentation of themselves in such a light might be said to have taken place in 1892.

In June 1892 some 20,000 people gathered in the great pavilion built specially for an anti-home rule protest, though used also by Dwight L. Moody for his evangelistic mission. Prayers were said by old Archbishop Knox, who had been a Bishop since 1849, and was the first primate to have been elected after disestablishment. The new Presbyterian Moderator, R. McCheyne Edgar, added his prayers. There were speeches from Thomas Andrews, Non-Subscribing Presbyterian, whose son was to become second Prime Minister of Northern Ireland, and from the Anglican Duke of Abercorn, whose son was to become the first Governor of the Province. Colonel Saunderson, fervent evangelical, Orangeman and landowner, had helped set up the meeting. The Revd R.R. Kane, prominent Belfast Orangeman and Anglican cleric, expressed the Orange and Protestant creed well when he spoke of the 'day when Irishmen, as well as Englishmen and Scotchmen and Welshmen, will be content to live and to work and to prosper under the aegis of an Imperial Constitution and parliament which secures the utmost civil and religious liberty for all and, with a strong hand, puts down arbitrary dictation, whether of the hired agitator or of a domineering clericalism'.[88]

Thus, at any rate in Ulster, Irish Protestantism, Unionism and Orangeism were, for many individuals, overlapping expressions of their religious and political and social beliefs. The subsequent decades cemented these beliefs, with the growing apprehensions about Catholic triumphalism, and the threat of a new home rule bill. In 1908, the Revd Alex Gallagher of Fountainville Street became the first Presbyterian Grand Chaplain to the Orange Order, and when the Covenant was drawn up in 1912 it was a Presbyterian, Thomas Sinclair, a one-time Liberal, who could claim authorship. Towards the end of the twentieth century, the historic identification of Orangeism with the Church of Ireland seemed to be proclaimed afresh in successive 'Drumcree' crises, centred on the historic parade by Portadown Orangemen to and from Drumcree Parish church, which led to the first formal attempts by the Church to distance itself from the Orange Institution. But in fact by 1991 it was reckoned that more Presbyterian ministers (13.5%) than Anglican (12.2%) were members of the Orange Order.[89]

Perhaps there is an alternative way of looking at the connection between Irish Protestantism and the Orange Institutions. The Orange Order, it has been suggested, was 'a pseudo-Church which brought together those who variously attended church and meeting. At its own meetings "in lodge" it provided ritual of a character missing both in the Presbyterian congregation and in the self-consciously Low church liturgy of the Church of Ireland.'[90] Certainly its rituals were as baffling to an outsider as Catholic rituals were to the average Protestant, and its orders and degrees and titles were as arcane as the orders and titles in the Vatican and the Religious Orders. Its art proclaimed vividly its beliefs, in a way that did not happen in Protestant Churches, but which was a vital part of Roman Catholicism. Banners depicted the heroes of the faith and of imperial history; biblical images abounded – 'David and Goliath, Naomi and Ruth, Moses, Samuel and the Three Wise Men' were among the not altogether predictable topics portrayed.[91] Symbols decorated the arches erected for 'the Twelfth': doves and five-pointed stars, square and compasses, red crosses, coffins, skulls and suns.[92] Marchers in Orange processions wore collarettes or sashes, gloves and bowlers, and carried pikes or swords or rolled umbrellas. So it is perhaps not fanciful to see in the Orange Order an outlet for Protestants whose churches and church services lacked the colour, spectacle and symbolism that the Roman

Catholics valued, but which – partly for that very reason – could not be adopted by Irish Protestants.

The rituals and fellowship which Protestants appreciated in the Orange Order could be found too in another institution, the Masonic Order. 'The exclusive Protestant fellowship of Masonry and the security provided by its internal charitable activities', writes Kurt Bowen, 'both played a part in it appeal, while some may have found that its complex rituals filled a need that was not being met by the rather plain liturgy of Protestantism.'[93] There were differences of course. Masonry was almost exclusively middle class; it was secretive; there were no great public displays which were so central to Orangeism. Nor was Masonry specifically Christian, and that would eventually become one of the objections to it. In the late eighteenth century, Masonry appears to have been plundered both by Roman Catholic and Protestant groups for ideas about organisation, symbols and ritual, and some of these found their way into to the developing Orange tradition. By the early twentieth century, many Irish Protestants were both Orangemen and Masons, and no doubt comforted in the knowledge that Freemasonry had been condemned on numerous occasions by the Papacy. In Britain its membership included church dignitaries and members of the Royal Family. In Ireland, Godfrey Day, Bishop of Ossory (1920–38) and briefly Archbishop of Armagh, was 'a keen and conscientious member of the Masonic Order', and presumably he was not the only member of the episcopal bench who 'saw in it a power making for public righteousness and for promoting brotherliness and charity'.[94] Masonry came under attack in the Free State in 1929 when Father Edward Cahill published his *Freemasonry and the Anti-Christian Movement*, attacking the world-wide conspiracy of Jews and Freemasons. Chancellor Kerr of Bainbridge said the attacks were 'due to ignorance, wilful and otherwise', and disputing the charge that it was 'irreligious' claimed that 'British Freemasonry is permeated with the reverence due to God'.[95] Kerr also gave his support to another organisation which appealed more to working men, the Royal Ancient Order of Buffaloes, which like the Masons was involved in 'works of mercy', but which unlike them – and to Kerr's regret – all too often met 'in public-house premises'.[96] In 1930 Archbishop D'Arcy preached at the dedication of a Masonic pillar in St Anne's Cathedral, in which a 'procession of the Craft, Royal Arch and High Knight Templars Masons' took part.[97] This evidence of the relationship between Protestantism and Freemasonry did not go unnoticed, and in 1935 Masonic halls, as well as Protestant churches, were the objects of arson attacks in the Free State as a reprisal for the sectarian violence in Belfast. The relationship was fading however. In the Church of England, Archbishop Fisher (Canterbury 1945–61) was 'probably ... the last senior ecclesiastic to be a Freemason'.[98] In 1977 an 'unprecedented' attack was made from a pulpit in Larne, when the Masons were called 'dangerous, unbiblical, un-scriptural and anti-God'.[99] That was by a Free Presbyterian Minister, who claimed that no communicant member of his church could be a Mason, though being an Orangeman was allowable. In the 1980s the Elim Church announced that no Mason would be admitted to membership. By then, it was becoming common even in the larger churches for questions to be asked about the compatibility of Masonry and Christianity. By the 1990s, questions were also being asked, but more tentatively, about the Orange Order and its links with the Protestant churches. Perhaps the age when Protestants needed colour and ritual in their lives was now past. Perhaps, like Masonry, Orangeism seemed something of an anachronism to many people. Perhaps membership of golf clubs and fitness gyms provided some of the sociability that had made Masonry and Orangeism attractive. Perhaps eating out, watching television, foreign holidays and shopping now gave the colour and the variety and the excitement that those organisations had offered. The 'Drumcree standoff' in the 1990s was however a reminder that the old loyalties and traditions had not been entirely forsaken.

4
Seeing and Believing, 1900–1965

> It is not often that the priest dilates upon the future happiness of good men, but when one eloquent curate, within my knowledge, expatiated with godly fervour on the splendours of the New Jerusalem of the Apocalypse to an old man dying in misery in a hovel near the village of Ballinamuck (the town of the pig), he only obtained this recognition of his efforts: 'Well, yer reverence, as ye're so intimate with th' Almighty, I wish you'd tell him from me, that if it's the same to him, I'd sooner stay in Ballinamuck'.
>
> *J.P. Mahaffy, 1912*[1]

Iconography and hymnography

'Service in the church on Derryvad hill was tiresome, dull, comfortless, in those terribly straight-backed pews', Sean Bullock remembered in his fictionalised autobiography of life in rural Fermanagh in the late nineteenth century.[2] 'Very dreary', was Lady Alice Howard's entry in her diary after a service in 1882; 'Such a dull old parson and the church nearly empty.'[3] A novelist took up the theme: '... a dismal, barn-like building so cold and damp that even in summertime it struck a chill through one. The floor is ill-paved, the plaster peeling off the walls, the cushions moth-eaten.'[4] Harold Nicolson had a different, but equally unspiritual, recollection of services in the 1890s when as a child he stayed with his kinsman, the Marquess of Dufferin and Ava. The huge party from the big house went to church in a procession of assorted vehicles carrying the marquess, the family, and the servants, and then made an 'entry into Bangor Church [which] had about it the solemnity of a State procession'.[5] Many years later, Nicholas Mansergh attended a service in St Anne's Cathedral, which 'is hideous, lacks proportion, style and beauty', as he noted in his diary, adding that he was 'bored to extinction' by the sermon.[6] And if church services were seen in this unspiritual and rather joyless light, so too was the Protestant Sunday. It was with modified rapture that the Presbyterian missionary, Matthew McCaul, remembered his boyhood Sundays in Londonderry in the 1890s. Coming from a respectable household – his father was a doctor, and an elder in Strand church – he observed a ritual which was widespread among middle-class Protestants. Sunday school was at 10.00 a.m. followed by church at 12.00 noon. Dinner at 1.30 p.m. was followed by a walk, and church again at 6.00 p.m. Tea was at 7.00 p.m.; hymns and prayers ended the day. Meanwhile, 'we were expected to read only "good books", that is, religious books'.[7] Sir Henry Hervey Bruce's grandson recalled Sundays at his grandfather's house at Downhill, which 'would have given even Calvin some hints in austerity for his government of Geneva', with a 'ban on everything except waiting for the next religious exercise'.[8] Dublin Presbyterians clamoured about 'the subject of Sabbath desecration in the city and suburbs',[9] and claimed that the 'Greatest Question for the Church today' was 'how the latter half of Sunday shall be spent by Christian people'.[10] Anglicans too worried about 'the passing away of Sunday as a day or rest',[11] and the Church of Ireland Conference in 1910 spent much time debating the 'Secularization of Sunday'. The Earl of Donoughmore lamented that modern transport facilities had 'engendered a spirit of restlessness which has been severely, but perhaps not unfairly, described as the modern idolatry of bustle'.[12] This concern about Sunday observance was not confined to Irish Protestants; in Britain too a growing concern led to the founding of the Imperial Sunday Alliance in 1908. But what gave the issue more piquancy in Ireland was the contrast between Protestant and Roman Catholic attitudes. The 'desecration of the Sabbath' had long been held to be characteristic of Catholics, and an added incentive to oppose home rule and attack the nationalists.[13] The different life-styles and beliefs of Catholics and Protestants were evident in their attitude to Sunday. They were evident also, of course, in their church buildings.

P.T. Forsyth, the great Congregationalist preacher, declared that a Protestant church 'must be primarily an auditorium'.[14] C.H. Spurgeon pronounced that 'every Baptist place should be Grecian, never Gothic'.[15] The reasons are obvious enough. Gothic architecture recalled the Middle Ages, when Roman rituals and doctrines were believed to have overlaid the Christian message with superstitious and idolatrous practices. A nonconformist church was the setting for worship, and at the heart of worship was the sermon. The plain and unadorned style of Irish Methodist and Presbyterian churches is 'best characterised by such useful words as "trim" and "neat", much employed by writers of gazetteers in Ulster'.[16] In the earlier part of the nineteenth century, when the Gothic style stood for the Anglo-Irish establishment, Presbyterians and Roman Catholics 'both built classical temples in conscious contrast to the established church's Gothic towers'.[17] Later in the century, 'all denominations had embraced some form of Gothic', though in Presbyterian churches it was usually so adapted that the auditorium was well lit and open, with room for galleries, and with a continued focus on the pulpit. And the listeners were not distracted by symbolism or art. Plain walls, and expanses of polished wood were the norm, and decorative items were kept to a minimum. Biblical texts were popular, but that was almost the only 'art' in the church. The Presbyterians made much use of the 'Burning Bush', the characteristic Presbyterian symbol since its adoption in Scotland in the seventeenth century, with its attendant inscription 'Ardens sed virens' (burning but living).[18] It was what the later twentieth century would call a recognisable 'corporate logo', and for most Presbyterian churches, that was the extent of the 'art'. There was an English nonconformist call for 'more general use of sacred pictures' on the grounds that they 'preach a better sermon sometimes than that from the pulpit', and 'are no more likely to be worshipped in our Protestant churches than gas-brackets are'.[19] But pictures were generally confined to church halls or Sunday school classrooms; Christ as the good shepherd, and William Holman Hunt's 'Behold, I stand at the door, and knock' were particularly popular. Within churches the only pictorial representations were likely to be in the stained glass windows. Not that there was many of those. In a recent book on the cathedrals of Ireland, of the thirty-one in the Church of Ireland, only four get a mention with reference to their stained glass: St Macartan's (Clogher), St Colman's (Cloyne), St Columb's (Derry) and the Cathedral of the Holy and Undivided Trinity (Downpatrick). And none of that glass is older than the nineteenth century.[20] The founding of the Arts and Crafts Society by the Earl of Mayo in 1894 saw a revival in this and other areas, much of it crossing the sectarian divide. Stained glass work by Harry Clarke, usually in the form of memorial windows, was commissioned for a number of Anglican churches in southern Ireland in the years immediately after World War I, and there is also a war memorial window of his design in Clontarf Presbyterian Church.[21] The work of Wilhelmina

Geddes is also to be found in Roman Catholic as well as Protestant churches; her first ecclesiastical commission was a window featuring the Angel of the Resurrection at St Ninnidh's Anglican Church at Inishmacsaint, Co. Fermanagh, in 1912. In 1916 she provided four 'Parables' windows for the Presbyterian Assembly building in Belfast, and in 1920 a splendid representation of 'The leaves of the tree were for the healing of the nations' at St John's (Anglican), Malone. Her 1929 'Christ blessing the little children' in Rosemary Street Presbyterian Church in Belfast was destroyed in an air raid in 1941. In all this – and more modern – stained glass in Protestant churches, the 'Protestant' nature of the work is evident. Saints or bishops might be represented, but rarely in the episcopal or eucharistic vestments common elsewhere. And as befits windows inspired by the Arts and Crafts movement, there is an abundance of animals and plants and maidens. A fine example might be Beatrice Elvery's pair of windows (1908–9) in the Anglican Church at Magheralin, Co. Down, 'depicting St Columbanus and St Gall, in which the main figures are accompanied by little angel girls who look as if they have strayed from a painting by Filippino Lippi, while the bird and animal life that inhabits these windows is a legacy of the Pre-Raphaelite approach'.[22] There was nothing here to disturb Protestant sensibilities. The Arts and Crafts movement had another effect. The re-opening of the Royal Irish School of Art Needlework by Lady Mayo in 1894 resulted in the spread within the Church of Ireland of richly embroidered altar frontals – there were particularly fine ones for Kildare Cathedral and St Patrick's, Dublin. Many would balk at calling them 'altar' frontals, but again the designs were calculated not to offend. Even the cross – the main Christian symbol – is absent, as it is from most stained glass and other work from the Arts and Crafts movement which found its way into Protestant churches.

So, visually, Irish Protestants did not get a great deal of stimulation from their churches, not even from the pre-eminent Christian symbol, the cross. It has been claimed that throughout the nineteenth century 'to put a cross on the gable of a church was to proclaim allegiance to Rome',[23] and the use of the cross was also forbidden within Anglican churches. Canon 36 of the church's constitution declared: 'There shall not be any cross, ornamental or otherwise, on the Communion Table, or on the covering thereof, nor shall a cross be erected or depicted on the wall or other structure behind the Communion Table, in any of the churches or other places of worship of the Church of Ireland.' That rule stood for almost a century, and other related rubrics, forbidding candles, incense, vestments and processions with banners, meant that the visual richness which by the beginning of the twentieth century was becoming widespread in the Church of England never infected its sister church. The Protestant Defence Association of the Church of Ireland carried on a continuous campaign to ensure that the purity was maintained, and 'high church' places like All Saints, Grangegorman, St Bartholomew's, and the TCD Divinity School (all in Dublin) were frequently the targets of their protests and litigation. Even in England, of course, there was widespread opposition to ritualistic practices more extreme and more common than anything in the Church of Ireland. The Public Worship Regulation Act of 1874, designed to curb excesses already far more serious than those prohibited by the Church of Ireland Canons, was a blunt instrument effectively overturned in 1906, after a Royal Commission found that 'modern thought and feeling are characterised by a care for ceremonial, a sense of dignity in worship, and an appreciation of the continuity of the Church, which were not similarly felt at the time when the law took its present shape'.[24] No such developments had taken place in Ireland. But why, in Ireland, such opposition to the use of that pre-eminent Christian symbol, the cross? T.C. O'Connor, Canon of Christ Church, Dublin, published a 77-page booklet in 1894 which set out the arguments.[25] He quotes verses written by the Scottish Presbyterian Horatius Bonar, author of such hymns as 'Fill thou my life, O Lord my God' and 'Rejoice and be glad! The redeemer hath come'. These particular verses focus on the kind of cross that might be worn about the neck:[26]

> Shall I call this glitering [*sic*] gem,
> Made for show and vanity;
> Shall I call this gaud a cross –
> Cross of him who died for me?
> Shall I deck myself with thee
> Awful cross of Calvary?
>
> Cross of man's device, I turn
> From thee, to Himself, my Lord;
> What can this symbolic gem
> Do for me? What peace afford?
> Shall I deck myself with thee,
> Awful cross of Calvary?

So, theologically, the objection was that no symbol of the cross could possibly replace, and might even detract from, the 'awful cross of

Calvary'. For most Irish Protestants however, the objection was simply that the cross used as a symbol was 'Romish', and therefore unacceptable. In 1895 Dr St George, Worshipful Master of Loyal Orange Lodge 152 (Lisburn) found himself on the carpet because of 'voting for the Cross in the Dublin Synod and other alleged Romanizing practices', and he was expelled for a year.[27] Forty years later, the Rector of St Bartholomew's, Belfast, was under attack for having accepted the gift of a communion table (suspiciously, 'one of the semi-altar type') which was adorned with 'IHS' carved in a quatrefoil, with additional ornamentation of 'foliage at the end of two cross bars', which took the form of a cross. He was accused of importing Romish symbolism.[28]

By the time of the St Bartholomew's affair, however, attitudes had shifted, albeit marginally. The Great War had done something to broaden the horizons of the men who fought on the western front, where many of them, like their English and Scottish comrades 'came across crucifixes, often for the first time in their lives'.[29] 'Crucifix Corner' was a map reference at the Somme. It was a Church of Scotland chaplain who reported that soldiers were impressed that they had often 'come across churches which had been shelled and broken to pieces, but one part remained, namely, the wall bearing the crucifix, the cross with a figure of the Saviour upon it. In the little shrines by the roadside the same thing happened.'[30] The poetry and art of the war were shot through with images of the dying Christ, the crucifixion, and self-sacrifice. Archbishop Bernard, preaching in Cambridge in 1917 – exactly two years after the death of his own son in battle – spoke of the soldiers' 'practical Christianity of the trenches', where 'deep down in these men's hearts is the great tradition of the cross. They are ready to die that others may live. That is the first lesson of the cross – only the *first* lesson; but this at least these men have learnt, and who shall say that it is not the greatest lesson of all?'[31] It would be hard to believe that such imagery did not have some effect on the men who fought, and on those who agonisingly watched. Even the Presbyterian newspaper carried an editorial, in 1915, on 'The Ministry of Angels'. Angels, it is true to say, were more acceptable in Protestant circles than representations of the cross, though they appeared but infrequently in stained glass or in cemeteries before the war. Yet now the editor could proclaim that 'we feel justified in believing that God revealed His power and protection in special ways to his servants. If some of the soldiers tell us with conviction that they saw a vision of Angels, we need not doubt their word.'[32] Thus the angel of Mons was an acceptable mystery, in a way that the story of the vision of the Blessed Virgin at Knock in 1879 certainly was not. And it was only much later, probably not until well after World War II, that the cross became an acceptable symbol on Irish Protestant literature and in churches in one form or other. But while there was no legislation to prevent non-Anglican churches from using the cross, the Anglicans still had Canon 36. An attempt to repeal it was made in 1929, led by Bishop Day of Ossory, who was not a ritualist. He had been educated in England, at Oakham and Pembroke and then Ridley Hall, Cambridge. There were no particularly high church influences in these institutions – perhaps the reverse. That he was still an Irish Anglican at heart was revealed in a discussion with fellow curates while he served in London's East End. The problem of church reunion would be solved, he declared to the amusement of his fellows, 'if we would only get to work and convert all the Roman Catholics to Anglicanism'.[33] But perhaps his English experience had broadened his mind. At the 1929 General Synod he (and Bishop Orr of Meath) introduced a Resolution, to be followed by legislation the following year, for the repeal of Canon 36. As it happened, the resolution was disallowed procedurally, even though the presiding officer, Archbishop Gregg, declared himself in sympathy with Day's motion. The following year Day proposed the appointing of a committee 'to consider whether any relief can be given top congregations which are dissatisfied with the provisions of the Canons dealing with Public Worship and the ornaments of churches'. The clergy said 'aye' by 83 to 62; the laity were overwhelmingly against, by 136 to 32.[34] Unfortunately for Day, this move came in the wake of a long-running ritualist controversy, involving the Revd W.C. Simpson, Vicar of St Barthlomew's, Dublin. Simpson was an English Anglo-Catholic who came to the Archbishop of Dublin's notice in 1921, when parishioners complained of candles on the holy table, the use of coloured stoles, the celebrant at holy communion taking an eastward position and making the sign of the cross, and the mixing of wine with water during that service. These were in fact five of the 'six points' (the other being the use of incense) regarded as indicators of 'high churchmanship' in early twentieth-century England.[35] Correspondence between Simpson and Archbishop Gregg, and complaints from parishioners and others, rumbled on until 1928, when the vicar was admonished and fined. A similar case arose in the early 1930s, against the Revd S.R.S. Colquhoun, incumbent since 1930 of St John's, Sandymount, Dublin. Petitioners in 1934 presented 21 charges against him, similar to those against Simpson. Matters were not finally resolved until 1939, when the vicar was permitted to retain a crucifix behind the pulpit, 'in the place which it had occupied for many years'.[36] Had it been a cross on the communion table, the order would have been to remove it! Other issues during the 1930s caught the attention of the Irish Church Union, with its aim of 'promotion and defence of the Reformed Faith of the Church of Ireland'. It was particularly active in the 1930s and 1940s, keeping vigilant watch, like the Protestant Defence Association in previous decades, to ensure that no elements of popery should creep into the Church of Ireland. Bishops were criticised for adopting the wearing of pectoral crosses; clergy were attacked for wearing short surplices, resembling the Roman cotta.[37] Clearly anything visually reminiscent of Roman Catholicism was to be eschewed. In St Kevin's in Dublin, 'regarded as the flagship of the evangelicals' in the 1950s, there was a notice in the vestry 'urging those taking services to refrain from wearing a cassock. A floor length surplice was the order of the day.'[38]

In 1964, thirteen years after Simpson resigned as Vicar of St Bartholomew's, and five years after Gregg retired as Primate, the General Synod passed legislation to allow a cross 'to be placed on the Communion Table or on the covering thereof or on the wall or other structure behind the same'.[39] Churches however are slow to change, and it was initially reported by manufacturers that 'orders have not been as high as expected'.[40] An ultra-Protestant news sheet nevertheless declared in 1969 that 'the practice has been sweeping through the denomination'.[41] Interestingly, the same issue of that newspaper carried the headline 'Ulster should support Rhodesia'! By the 1990s, all the bishops (as pictured in the *Church of Ireland Directory*) were wearing pectoral crosses. Coloured stoles, and even the occasional cope and mitre, were seen among the episcopate. In the last decade of the twentieth century, the bishops of the Church of Ireland were beginning to adopt practices that were innovations among English bishops in the first decade of the century.

If Irish Protestants rejected the use of the symbol of the cross, there was nevertheless a verbal imagery in place of the visual. Ernest Rattenbury once described Isaac Watt's great hymn, 'When I survey the wondrous cross', as 'a verbal crucifix, built up of carven words'.[42] Irish Protestants might not want to see a cross in their churches, but they could, like Watts, be found 'surveying the whole realm of Nature and finding at the centre of it its crucified and dying Creator'.[43] Protestants might find Catholic devotion to the sacred heart offensive, and the representations of it in the pictures in Catholic homes utterly alien. The idea around in the south in the early 1920s that a statue of the sacred heart be erected atop Nelson's Pillar in Dublin, or alternatively featured on the flag of the new state, must have seemed grotesque to them.[44] Yet verbal representations of the same thing were acceptable, and the imagery of blood was common in Protestant hymn books. Particularly in the Chapman-Alexander or the Moody and Sankey mission hymn books, the imagery is vivid indeed:

Heir of salvation, purchase of God;
Born of His Spirit, washed in his blood.

He died of a broken heart for you,
He died of a broken heart.

There is a fountain filled with blood, drawn from Emmanuel's veins,
And sinners washed beneath that flood lose all their guilty stains.

There is power, power, wonder-working power,
In the precious blood of the lamb.

Interestingly though, lest there be any doubt that the imagery is historical and verbal, the Chapman-Alexander hymn book amends the last verse of 'Abide with me', written by the Irishman, Henry Francis Lyte, and possibly the most popular hymn throughout the British Empire in the late nineteenth and early twentieth centuries. Thus 'Hold thou thy cross before my closing eyes' becomes 'Be thou thyself before my closing eyes'. And as far as the imagery of blood was concerned, it was an unusual Methodist minister who attacked the hymn book for its frequent references to Christ's blood; oddly, the examples he chose were hymns by Charles Wesley and translations by John Wesley. 'The picture galleries in Roman Catholic countries', he wrote, 'are crowded ad nauseam, with paintings of our Lord upon the Cross, or as taken down from the Cross – a ghastly spectacle; but the Protestant Churches rejoice in a living and exalted Saviour.'[45] While most Protestants would have agreed with his last comment, few Methodists would have committed the 'heresy' of attacking the Wesleys' hymns.

If Protestants could not feast their eyes on vestments, artefacts or symbolism in their churches, their imaginations were certainly fed by the imagery in their hymns. Since the middle of the nineteenth century, there had been a great outpouring throughout the United Kingdom of new hymns and new tunes. The Victorians had an 'enormous enthusiasm for hymns', which brought together, as Ian Bradley has pointed out, many of the values of that age – 'high moral seriousness, sentimentality, populism, evangelical fervour, doubt-laden faith, high literary culture, scholarly antiquarianism, anguished optimism and a strenuous sense of duty'.[46] The Methodists were – among the larger denominations – the most enthusiastic singers, in possession (after 1904) of the most comprehensive hymn book. 'Methodism was born in song', as its preface and that of its (1933) successor proclaimed. Irish Methodists in the twentieth century may not have been to the fore in 'contributing to the creative arts', even though two of Ireland's foremost poets, John Hewitt and Seamus O'Sullivan, had Methodist upbringings.[47] But they could review their hymns and say with John Wesley: 'Poetry thus keeps its place as the handmaid of Piety.'[48] Along with Wesley's *Forty Four Sermons*, their hymn book set out their doctrinal teachings which, almost by osmosis, Methodists absorbed week by week. The Presbyterians were slower to adopt hymn singing, being wedded to their metrical psalms, opposed to the use of organs in church, and often reliant – even in the singing of these – on the services of a precentor. An 1870s survey of Presbyterian churches revealed that two-thirds employed a precentor, under half had no choir, and a 'large number' had no music books. The 1870s also saw the beginnings of the controversy over the use of organs, opposed by the Purity of Worship Association, founded in 1875. The General Assembly forbade the use of organs in 1882. This edict was not formally repealed, but when in 1892 the Assembly was told that Rathgar and Dundela were using organs, it voted to pass from the question. By 1917, at least half the churches had acquired an organ. Meanwhile in 1898 the first edition of a Church Hymnary was sanctioned, though for some 20 years thereafter some congregations refused to use it,[49] and even after World War II there were congregations who sang only psalms, without accompaniment.[50] Not that hymns were rejected altogether. Ethel Corkey found this puzzling: her father-in-law was fond of hymns, but would not sing them on a Sunday.[51] Not everyone was as liberal even as this. When the Revd W.F. Marshall announced a hymn at an Orange service at Auchnacloy, there were objections voiced, but he enjoyed the reaction when it turned out to be the Old Hundredth, which was acceptable to the Presbyterian traditionalist objector.[52] In 1927 a revised edition of the Church Hymnary was published, and a third edition in 1973. Its publication was accompanied by 'prolonged birth-pangs'[53] – as is the case with most new hymn books. By 1990, there was a feeling that the singing of psalms might be in decline – partly because the new hymn book contained sufficient psalms (57) to obviate the necessity of purchasing a separate psalter as well.[54]

The Anglican tradition was also less geared to congregational hymn singing than the Methodist, and its choirs needed training in the singing not only of hymns but of canticles and psalms. Few churches had choirs, and as one old clergyman remembered – recalling his ministry in the mid-nineteenth century – the Parish Clerk 'had a monopoly of the praises of the sanctuary. His usual preliminary notice was, "let us all stand up and sing", in response to which the people did not stand up, neither did they sing.'[55] Later in the century the Revd W.J. Murray felt that hymn singing was vital to more attractive services, and suggested that the answer would be 'if possible, to have the whole congregation a choir, and the song of praise ascending from all ...',[56] which suggests that there was still a problem in achieving that. Progress (difficult to define, and to trace) had been made by 1910, when 15 popular hymns were printed for the use of the Church of Ireland Conference, a third of them by the Wesleys or Watts. And when Joyce Collis converted to Rome in 1927 from a prosperous Anglican family in Killiney, she remembered thereafter how 'I certainly missed some of the beautiful hymns that I loved in the *Ancient and Modern Hymnal*, and the active part that is taken by the congregation in Protestant services.'[57] Even at that stage, church organists voiced a concern that 'Irish congregations are too shy to join wholeheartedly in congregational singing.'[58] But the Anglicans too had another dimension to their music which was unique to them: a long tradition of cathedral worship, with successions of organists going back to the sixteenth century (St Patrick's and Christ Church, Dublin) or the seventeenth century (Armagh, Cloyne, Cork, Derry, Kilkenny, Limerick). But in this, as in other ways, they were burdened with all too many cathedrals, under-endowed, small, and costly to maintain or restore. The 'great Anglican tradition of Choral Evensong ... was a Victorian invention',[59] and one that given local constraints could achieve only limited success in Ireland. Naturally the tradition was particularly cherished in Dublin, though even here there was the problem of having two cathedrals close to one another, with limited endowments and a limited 'market' for their product. Each cathedral managed to maintain its choir school until 1972, when the Christ Church one closed, despite the best efforts of Dean Griffen of St Patrick's to arrange some kind of merger.[60] By 1981, St Patrick's was able to boast of a new, co-educational choir school – the only one left in Ireland – and was by then 'the only cathedral in these islands at which daily matins is still sung, and ... the sole cathedral in Ireland at which there is a choral evensong daily'.[61] St Anne's cathedral, despite being consecrated in 1904 at the height of Belfast's prosperity, never acquired its own choir school, and never developed a tradition of daily services, though it did break ground by incorporating ladies in its choir. This may have been because of a personal preference by the organist C.J. Brennan, who played at the consecration service in 1908 and was – astonishingly – to retain the post until 1964. The cathedrals of Ireland did however feed into the wider Anglican church music tradition. It was two Irish boys, C.V. Stanford and Charles Wood, who became the most noted late Victorian composers of church music.

Joyce Collis missed Anglican hymn singing. Something else she might have missed from the poetry of her former church was the common practice – probably until after World War II – of 'learning by heart the collect of the Sunday, the whole of the catechism and many of the psalms'.[62] George Otto Simms, at preparatory school in England in the early 1920s, had to learn the collect of the day each Sunday morning, with the result that 'sixty years later the Cranmerian phrases roll[ed] off his tongue with consummate ease'.[63] As with Methodists through their hymns, and Presbyterians through their psalms, Anglicans through their Prayer Book absorbed and internalised both the rhythms of the language and also the theology of its meaning. But of course in this highly verbal culture, the Bible was the most important source. Sean Bullock's recall of the 'dull, featureless' church of his Fermanagh boyhood was not the only religious memory that affected him. He mused on his knowledge of words from Isaiah, from Ecclesiastes, from John's Gospel, and he asked:[64]

> Is it a small thing that a boy shall grow to manhood, his life resonant with such sentences, hundred and hundreds of them? Apart from their inner significance – which possibly he may never feel – the sublimity of them, their beauty, their music, must be like the everlasting arms sustaining him, a sure refuge and defence, now when the heart leaps, now when grief or pain assails, and he lies in the dark having little more than the hoards of memory to give comfort. Even in every-day times of health and work, life is the richer for such stores.

In the lives of Irish Protestants such resonances played an important part. So too in their deaths. Protestantism lacked the pre- and post-death rituals so important to Roman Catholics. The sacrament of extreme unction, the requiem mass and the wake gave Irish Catholics an extended structure of comfort and support for death and bereavement. For Protestants, there were the words of scripture and of hymns. When the Revd William Moutray, old and infirm, faced the new century, he recorded in his diary:[65]

> [3 February 1900] No-one called today; my nights are nearly sleepless; but I am happy in my mind and in the prospect of never ending joy in that bright world above.

> [9 February 1900] I had no visitors or letters today. My state of weakness and pain continues and my only happiness is to think of the time when I will be in heaven and this hope will support me to the end, and the joy of being with my dear Saviour will make up for all the pain I now endure.

> [17 April 1902] I read God's word constantly and I pray for faith to believe in it, and to hope I will soon be in heaven, that happy home where all the redeemed will meet, never to part more.

Dean Ovenden of Christ Church commented to a friend in 1920, on hearing 'Lead, kindly light': 'When I hear that hymn so played I feel glad I am an old man and a dying man.' And, as his friend noted, when he named his favourite hymns, they were 'all similar in tone, bright with glimpses of the hereafter'.[66] The words of the Bible and the hymns of Irish Protestantism helped to shape their perceptions not only of the here-and-now, but also of the hereafter. Death was all too common, and hymns helped people cope. Mrs Alexander's children's hymn on the subject – macabre to the modern reader – must have performed a useful function in an age when death was not followed by the ministrations of professional counsellors:

> Within the churchyard, side by side,
> Are many long low graves;
> And some have stones set over them,
> On some the green grass waves.
>
> Full many a little Christian child,
> Woman, and man, lies there;
> And we pass by them every time
> When we go in to prayer.

The Fermanagh farmer who wrote to his emigrant brother in Australia telling of their father's death was one of those Methodists, common in those parts, who had Anglican roots as well. Those influences, and a sincere faith, are evident in a letter that combines lack of accuracy in spelling and grammar with a deep appreciation of the faith, and the rhythms of its verbal expression in hymn or Bible or Prayer Book.[67]

> I never witnessed sutch a seine [scene] to see him Praying to God to his last moments. We have every right to be thankful to God when It was his will to take him too see him pass triumphant home to Glory. And our friends and Neighbours told us that we had no right to be sorry. But we must bear the trobles and trials of this world ... My Dear Brother I pray to God that us and yours may be as well prepared to meet Our God as our Father was. He said to me a short time before he died that heaven is worth a Contending for. I hope we shall all meet him there.

Preaching

The sermon was the highlight of a service. The architectural arrangements in almost all non-Anglican churches gave the pulpit pre-eminence. There were some exceptions. The Methodist Carlisle Memorial and Fisherwick Presbyterian for instance mirrored the layout of an Anglican church, with the communion table as the focus, but they were unusual in the early part of the century. Most often the pulpit loomed large, and usually centrally, proclaiming the centrality of the preaching of the Word. There, and in gospel halls and in the tents erected for special missions, the sermon was a great draw. During the nineteenth century 'sermon-tasting was a favourite diversion that also counted as a duty and was thus doubly attractive to the Victorians'.[68] For the first few decades of the twentieth century this was true in Irish Protestantism, and indeed the tradition of sermons and evangelistic missions continued to flourish thereafter – particularly in Northern Ireland – in a way that was more American than British. The period at least until World War I was an age of great political as well as ecclesiastical oratory. In a society not yet saturated with the sounds from radio and television, there seems to have been a far greater desire for, and ability to listen to, extended verbal argument and exhortation. The prolixity, as it now appears, of speeches and sermons seems actually to have been part of their appeal. Sermons were educational and emotive; they were entertaining and often arresting. And they could be very long. It is hard to believe that when the Presbyterian Robert Watts preached without notes in May Street Church, Belfast, on the subject of 'Millenarianism', he was able to keep going for an hour and a half, and interest 'never flagged for a moment'.[69] The Revd John Waddell

(Fisherwick 1920–45) claimed that if 'a sermon is worth preaching it is worth listening to for twenty-five minutes or half an hour', and this seems to have been a more standard time than Robert Watts' hour and a half. The Anglicans were less keen on lengthy sermons. In the late nineteenth century the Revd J.W. Murray was claiming 'that shortness is an excellent thing in a sermon', and that its power of doing good was 'in inverse proportion to its length'.[70] Indeed at Christ Church Cathedral in Dublin the chapter suggested '15 minutes as the maximum length of a sermon'.[71] Even so, Anglicans valued sermons: when Canon F.W. Mervyn died in 1933, after a ministry of 52 years, it was proudly claimed that 'he had preached 4,673 sermons in various parts of Ireland, and he kept a record of them all'.[72] Indeed it was Anglican preachers who sometimes gained the plaudits for sermons or addresses which gripped their listeners. In any list of the greatest preachers in Victorian times, two Irish Anglicans would probably feature: William Connor Magee (who moved to England to be Bishop of Peterborough and – briefly in 1891 – Archbishop of York) and Archbishop Alexander of Armagh, 'the golden voice of Anglican preachers', as Shane Leslie called him.[73] By the beginning of the twentieth century Magee was dead, and Alexander was old and ailing. But the Irish could still hear great Anglican sermons. When Bishop Moule of Durham addressed the Church of Ireland Conference men's meeting in 1910, we are told that the entire audience was 'spell-bound' and then 'rose and cheered him to the echo'.[74] When he visited Ireland again in 1914, he had a similar effect after speaking in Dublin.[75] In the post-war era, one of the most popular and effective Anglican preachers was – at least by birth – an Irishman, G.A. Studdert-Kennedy, popular with the troops during the war as 'Woodbine Willie'. He has been described as standing 'in the true line of succession' of his fellow Irishman and leading high churchman in the late nineteenth century, Father Dolling.[76] Like Dolling, he would not have been at home in the Church of Ireland. Like him, he was something of a Christian Socialist, and a populist preacher. Like him, he could speak to people at the upper end of the social scale as well. When he preached in the Chapel Royal in Dublin in 1921, Lord Oranmore made a point of going to hear him.[77] Ten years later a future Archbishop of Canterbury, William Temple, 'almost mesmerized' the undergraduates at TCD by his evening addresses.[78] Future generations of students were to be less impressionable, an indication both of the decline in the power of the sermon, and of the greater scepticism about religion and about figures of authority which from the 1960s on were characteristics of university life. There was some fear that when Billy Graham visited Queen's, Belfast in

1972 the students might 'break into catcalls and jeers at any moment'; in the event, 'they could not have been more attentive'.[79]

The Presbyterian founder of the 'Catch-my-Pal' temperance movement, R.J. Patterson, was another preacher whose reputation early in the century indicates a power of the spoken word scarcely credible to later generations. Talking to men for 'a couple of hours he could keep an audience alternating between laughter and tears'.[80] D.L. Moody's visit to Belfast in 1892 had given fresh impetus to the centrality of the sermon, and the tradition of 'gospel' preachers was to thrive in Ireland throughout the twentieth century. The most influential Irish preacher in the 1920s was W.P. Nicholson, who was converted in 1899, the year of Moody's death. Nicholson studied at the Bible Training Institute in Glasgow, founded as a result of the Moody missions, and had a preaching apprenticeship with Chapman and Alexander in Australia and the USA. John Chapman had worked with Moody, and Charles Alexander had teamed up with Reuben Torrey (playing the Ira D. Sankey to Torrey's Moody) in highly successful missions, including one in Ireland in 1903. Nicholson had a great talent for talking to working men or, in the Belfast of the 1920s, men out of work. His language was rough and often humorous. Listening to the singing on one occasion, he interrupted: 'It reminds me of Paddy in a high-class restaurant who had all the courses and then said, "Your samples were good, but bring my dinner now".' Preparing to speak, he once said: 'Get your nose wiped ma'am and get your glasses on. You bald headed fellows get ready.'[81] Defending Nicholson's use of 'crude language' at a Dublin mission, the Revd T.C. Hammond wrote: 'Do they suppose the truth of Christianity always requires genteel drawing room talk?'[82] At a 1926 mission in Cambridge, 100 undergraduates professed conversion after Nicholson had come in as 'an incongruous last minute substitute'.[83] His Irish missions were welcomed across the churches. The Revd John Redmond, in his Ballymacarrett parish newsletter of March 1923, noted the success of a recent mission held in the nearby Methodist church: 'Men marched in hundreds from the shipyards to the services. On one occasion so great was the pressure of men to get through the Church gate that the top of one of the pillars was pushed off!'[84] One young Presbyterian remembered a similar march by 'singing ranks of shipyard men' processing from the station to the Presbyterian Church in Ballymena, singing Alexander hymn book favourites.[85] The Presbyterian Belfast City Mission reported, after these missions and those of Le Marechale (Salvation Army William Booth's daughter): 'It was only natural that the influence of these campaigns should be felt throughout the whole community, and the

City Mission shared in the benefit of an increase of zeal and a more earnest inquiry after the things that pertained unto eternal life.'[86] In the wake of these missions in the early 1920s, there were doubts about the efficacy of modern preaching; after all, most clergy could scarcely emulate the rhetorical successes of the few. 'Is the power of the pulpit decaying?', the Presbyterian newspaper asked in 1927.[87] The editor was not quite sure. On the one hand, he noted, the educational standard of congregations was higher than in the past, making them perhaps more critical. On the other, it was all too easy to invest 'the pulpit of the past with a halo which careful investigation may readily dispel'. His conclusion was that 'the pulpit today is exercising a wider and deeper moral and spiritual influence than at any date in the past' – itself an assertion which 'careful investigation' was unlikely either to dispel or verify. St John Ervine, speaking at the Portadown Music Festival in the following year, had no doubt that the younger clergy lacked rhetorical skills, and 'could not deliver the Gospel without moaning and droning and almost whining over it'.[88] 'The parsonic voice must be abandoned', he declaimed, though this criticism might be applied to at least some clergy at any stage in history.

Meanwhile, the 'apostolic succession' of evangelical preachers was maintained in Ireland. Over two decades after Nicholson's great 1920s missions (while he was still preaching but was less high profile), his influence helped to form the style and message of another noted evangelist, Ian Paisley. Paisley recalls how, a week after his ordination in 1946, Nicholson attended his first sermon at his Ravenhill Road mission hall. He addressed Paisley and the congregation: 'I have one prayer for this young man, that God will give him a tongue like an old cow. Young man, go into a butcher's shop and try and run your hand along a cow's tongue; its as sharp as a file. Please God this man will have a tongue that shall be as sharp as a file in the heart of the enemies of the king.' As his biographer tells us, 'Paisley believes that God answered that prayer.'[89] Few would disagree as to the effect, if not the cause. Paisley continued the grand preaching tradition, and in 1969 his 'Martyrs' Memorial' Church was opened with a seating capacity of over 2,000. Whether, as was suggested in the 1990s, the numbers dwindled dramatically after the packed congregations in the previous two decades, the tradition of preaching to huge numbers was still alive in Northern Ireland. In 1994 the non-denominational Whitewell Metropolitan Tabernacle was opened, with a seating capacity of over 4,000. It is light and airy; it has foyers and fountains, chandeliers and murals – more like an American cathedral of light than an Ulster gospel hall or church. Its pastor, James McConnell, claims Charles Haddon Spurgeon as a boyhood hero.[90] Over a century after Spurgeon's great ministry of preaching in the Metropolitan Tabernacle in London, Belfast's Metropolitan Tabernacle continues the tradition – in a very different world. The greatest exponent of that tradition in the twentieth century – Billy Graham – visited Belfast in 1972.[91] Prevented, by security reasons, from holding the traditional 'Crusade', he walked and talked and prayed along the Falls and Shankill roads. During his walkabout he 'heard a mighty explosion' and 'soon came upon the dreadful sight: bodies and pieces of bodies blown apart by a bomb'. He and his friends rendered what assistance they could. He preached in a church; he met the Governor, Lord Grey; he spoke to Catholic and Protestant students and chaplains at Queen's; he met with Cardinal Conway. In the Republic, he held meetings at the Royal Dublin Concert Hall and at the Jesuit Headquarters, Milltown Park, where he addressed clergy from all the churches. He met a Roman Catholic priest who 'told how he had come to Christ through reading my book *Peace with God*'. He talked with the leader of the Official IRA. So much of Ireland's religious and political history was encapsulated in that visit. To the 'verbal crucifix, built up of carven words', to the sermons of preachers, must be added their example, their lives, their activities as much outside the pulpit as in it.

Appearances

Clergy were respected in Ireland. In 1873 a new young Methodist minister wrote, in astonishment: 'It is amazing what influence anyone who bears the stamp of the Methodist Conference has over the people at large. I find myself an authority on any matter.'[92] Almost a century later a survey of Ulster young people's attitudes revealed that the clergyman was 'still one of the most influential and best-known persons in the community', 'well known' to between 83% and 97% of people from varied backgrounds.[93] Whether in the 1870s or the 1960s, Irish Protestantism existed in small communities, conservative face-to-face societies. The chance of being anonymous hardly existed, even in the cities. This affected the lives of the clergy, who were subject to more scrutiny than most lay people. However by the end of the century the Dean of St Patrick's Cathedral, Dublin, could claim that there had been a 'lowering of esteem for all clergy', whose standing had 'never been been so low as it is in Ireland today'.[94] In this, as in other ways, the pace of change had quickened in the last quarter of the twentieth century.

For much of the century dress was an indicator, in general, of social class. In the years before World War I, it can be seen in photographs: the social distinctions marked out by the type of male headgear, the top hat, the bowler, the cloth cap. And the clothing of the laity might even indicate something about their denominational affiliation. In the 1940s, there were 'people who were dressed in such a way that they felt more at home in a mission hall'. Poorly dressed people would certainly feel less conspicuous in a 'tin tabernacle', or in a city mission, than they would in a normal church. There were others, as one Presbyterian remembered, who were 'dressed in a way that made others feel they should have gone to the Established [*sic*] Church'.[95] In the last quarter of the century, such dress distinctions had all but died out, as comfort and informality affected even the church-goers of Protestant Ulster. But for clergy, as for policemen, dress also indicated their job. It was not regarded as incongruous in Edwardian Belfast that when visiting poor parishioners a young clergyman should wear the top hat and frock coat of his trade; it marked out his social superiority, but it was also a badge of his calling.[96] T.M. Johnstone, as a young trainee before World War I, was told by the minister of a church where he turned up to preach that he should not 'go out on supply without clericals' – meaning the clerical collar.[97] A young Methodist probationer found that 'people in the country liked us to go to them as young ministers, so most of us put on a clerical collar when we went to preach'.[98] By that time, the clerical collar or dog collar (or 'Roman' collar, but not so-called by Irish Protestants) had become the almost universal clerical dress for the bigger denominations. Older clergy might still – as was the case of the clergyman who told Johnstone to wear 'clericals' – prefer the white tie which had been common until at least the mid-nineteenth century. J.P. Mahaffy, Provost of TCD, after the Great War, 'wished to keep aloof from a new kind of clericalism'; the dog collar did not 'appeal to a man of his protestant background and independent intellect'.[99] However most clergy more 'Protestant' than he were prepared to wear the uniform. Anglicans, Methodists and Presbyterians alike wore the dog collar, and often the frock coat, until well into the second half of the century. By the 1970s, the 'dog collar' might be of the slip-in variety, and the traditional black accompaniment was under threat from 'vestocks of every hue'.[100] At the same time, however, the abandonment of the dog collar altogether by many clergy was becoming more prevalent. In 1969, there was a call for Anglican bishops and deans to abandon their traditional 'uniform' of gaiters and top hats[101] – and within a decade those quintessentially Anglican garments had been almost entirely abandoned both in England and Ireland. But the clerical 'uniform', specifically the clerical collar, was worn throughout most of the century, and ensured that the clergy were visible. This had implications for their behaviour. But their dress also indicates that the clergy, like other members of society, partook of the changing fashions round about them. This was true of the clothes they wore; it was also true of their personal appearance. Take a photograph of the hundred clergy at the Methodist Conference in 1886; every one of them had 'beards, whiskers and moustaches of various types'.[102] Look at a picture of the Belfast Town Mission officers in 1897; all but two of the twenty-two sport beards or moustaches; the other two (it is difficult to tell) may have sideboards.[103] Facial hair denoted manliness; absence of it could (though not invariably) denote effeminacy, often a code word in the nineteenth century for homosexuality. The Methodist J.H. Rigg, writing in 1895, 'made much of the "characteristically feminine" mind and temperament of Newman and the lack of virility among most of his disciples', while their Anglo-Catholic successors at the end of the century were described by Protestant militants as 'effete, decadent and lacking in manly qualities'.[104] In Ireland such criticism could more easily be directed at the Roman Catholic Church with what Sean O'Faolain called 'its effeminate ways'.[105] But fashions change. In the post-war world facial hair was no longer *de rigueur*. Of the twenty-nine Belfast Town Mission staff in 1925, only one elderly man sports a beard; three or four have small moustaches.[106] Of the nineteen men in a photograph of the installation of the Revd Robert Caldwell at 1st Carrick in 1940, all are entirely clean shaven except for one old cleric with a grey moustache.[107] Clergy and church workers, more than they might have liked to admit, were followers of fashion.

Should clergy follow the fashion for smoking? This question had wider implications, for it raised the issue of whether Christians should smoke; should smoking be regarded as a sin of the flesh? St Paul had condemned fornication, impurity, indecency, drinking bouts, orgies and such like.[108] He said that 'those who behave in such ways will never inherit the Kingdom of God', and Protestants in particular (and certainly the more puritanical Protestant groups) had added to that list. Betting, gambling and drinking (and not just drunkenness), might imperil the soul. Might smoking do the same? The Methodists and smaller denominations tended to take the firmest line. One writer in 1894 noted how 'the [tobacco] evil is becoming daily more widespread', but felt that Irish Methodist clergy 'are to be congratulated, for our clerical ranks are almost untouched; but the same cannot be said of the Church across the channel, where Episcopalian and Dissenter alike smoke as naturally as they preach'.[109] The 1905 Conference expressed its 'strong disapproval of the habit' of juvenile smoking, though it was not until 1962 that the medical evidence was such as to support a wider official disapproval. In that year, the Conference asked 'those who smoke to consider their position' as a result of the report of the Royal College of Physicians.[110] Presbyterians took a less prescriptive stance. Smoking evidently took place at some ordination dinners, a practice attacked in the middle years of the century as producing 'an entirely wrong atmosphere' even though the critic emphasised: 'I am not saying that smoking is a sin or that smokers cannot be good men.'[111] Not everyone was prepared to make that caveat. The *Irish Presbyterian* opened the new century expressing sentiments that would be far more widely endorsed a century later: that a minister 'with the powerful and unpleasant odour of tobacco on his breath' was hardly an appropriate visitor at a sick bed.[112] To the question 'Is the Use of Tobacco injurious?', medical and scientific reasons were adduced to give the answer 'yes'; but more than medically, 'cigarette smoking blunts the whole moral nature', with an especially bad effect on schoolboys.[113] Anglicans, however, tended to take a more relaxed view. The Revd J.O. Hannay, writing in 1934, noted that – 'odd as it may seem now' – during his early days in the ministry, in the 1880s, it 'was supposed that a man could not be a good parson if he smoked a pipe', and that his father 'had to give up smoking when he was ordained' (in 1859).[114] The editors of *The Warden* – not the most liberal of papers – poked fun at the Christian Endeavour ladies of America who appealed to ex-President Roosevelt's daughter 'to cease from her alleged practice of smoking cigarettes', and commented that 'these good people should mind their own business'.[115] Indeed a few years earlier they had welcomed the possibility that 'Ireland may yet become known as a tobacco-growing country', following reports of this as a possible new rural industy.[116] Bishops smoked. Godfrey Day (Ossory 1920–38; Armagh 1938–9) was an inveterate pipe smoker who observed the Lenten fast by complete abstinence from the use of tobacco – 'a very severe form of discipline', as his biographer records.[117] It was an Anglican Rector's wife in the 1930s who was remembered as 'welcoming us warmly to the rectory in Co Longford, cigarette in mouth and a cairn terrier under either arm'.[118] But then, she was an Englishwoman. In 1998, when the scholarly Henry McAdoo (Archbishop of Dublin 1977–85) died, his obituary affectionately pictured him with pipe in mouth – 'where it was for most of his waking hours'.[119] Not a tribute one could imagine being paid to a Presbyterian or Methodist leader. The issue of smoking is representative. It reveals something about the clergy and their respective denominations. The Methodists – and other smaller churches – were the most puritanical, and the most prescriptive. Their attitude to smoking is paralleled also in their attitude to drinking. The Church of Ireland was far more *laissez-faire*. The Presbyterians fell somewhere in between.

Should clergy go to the theatre? Here was a similar problem, or indeed more so, for it was a public act. J.O. Hannay again remembered that his father never went to the theatre himself, but did not object to his son going 'as unostentatiously as possible, lest I should create a scandal'.[120] He also claims that when Belfast's Theatre Royal burned down, the Ulster Hall was adopted as an alternative venue, and 'even Presbyterian ministers flocked to see plays!' Certainly the daughter of the great English Methodist, Hugh Price Hughes, recalled that although her father loved the idea of 'witnessing fine plays finely performed', he could not go to the theatre, because of the 'stage problem'. And so despite 'all his love of drama in its true sense, he was deeply aware of what he considered its seamy side'.[121] The Presbyterian newspaper in 1901 was able to give its readers 14 reasons why Christians should not go to the theatre.[122] In the decades after the eighteenth-century revival, evangelicals had attacked the 'frivolity' of the theatre, and more recently its associations with stage door Johnnies and musical hall vulgarity made it an object of suspicion among the serious minded. 'Clergymen in 1901, though by no means all, were willing to go to the theatre', says Owen Chadwick;[123] this was less true in Ireland. And again, Methodist and Presbyterian clergy – and certainly those of the Baptist and smaller churches – were more likely to feel inhibited than their Anglican brethren. At the Keswick Convention – much patronised by Irish and English evangelicals – in 1930, there was a declaration that 'God was calling them to an utter separation from everything that was questionable.'[124] Wireless, theatre and cinema were all listed as noxious influences. But by 1947 even the Inter-Varsity Fellowship in its profoundly conservative evangelical newsletter was urging ministers to drop blanket condemnations of cinema, theatre and tobacco: 'They were erecting false barriers that hampered the gospel.'[125] In Ireland the cinema, rather than the theatre, was seen in the inter-war years as a sinister influence, not least because it was so widely available and affordable, unlike the theatre. This was an issue that crossed the sectarian divide; Catholic priests were outspoken in their condemnations of the cinema, though seldom of smoking or gambling. As regards the cinema, clergymen had to be careful, if they went at all, as to which films they viewed – that was the price of being significant and respected figures in the community. Remembering his own early days as an Anglican minister in the middle of the century, Herbert O'Driscoll recalls that 'the confidence of the clergy beginning their ministry came from the realization that the society they were entering affirmed and accepted their role without question'.[126] To be a cleric then was 'regarded not only as quite normal, but even as admirable'. The changes in society itself – as it became more mobile, more affluent, more tolerant, less church-going – were bound to affect the public persona of the clergyman. Protestants were fond of quoting 1 Thessalonians 5, 22: 'Abstain from all appearance of evil.' Paul's advice meant that a minister, in particular, must be seen to 'to keep himself unspotted from the world' (James 1, 27). But appearances rely heavily upon prevailing fashions, and perceptions of goodness and purity depend, in part, on current fashions and prejudices. David Corkey, the Presbyterian minister, whose BB activities have been mentioned, gave himself, we are told in his wife's biography of him, enthusiastically 'to work amongst the young, especially amongst the boys'. 'The girls', she adds, 'felt a little "out of it".'[127] There is a photograph in her book, showing him at a BB camp in 1911. He is lying with his head on a boy's shoulder; another boy lies alongside, his arm draped over Corkey, holding his hand. They are all staring at the camera, eyes half-closed against the sun.[128] It is a photograph that could not possibly have appeared in a biography in the last quarter of the century. Thus do perceptions change.

5
Confessional States, 1922–1965

> Post-1921, each part of Ireland has had in varying measures the features of a confessional state, and this characteristic did not greatly diminish with the passage of time.
>
> *T.P. McCaughey, 1993*

The churches in a Protestant state

The Irish churches could not enter the post-war era hoping for a return to 'normalcy'. The guns had fallen silent on the western front, but not in Ireland. Two and a half years after the Armistice, the Primate of All Ireland opened the General Synod lamenting that the 'things that had happened in their country during the past year had been so terrible, so disastrous, so fateful in relation to the social and moral life of the whole community, as to be paralysing'.[1] A year later, he scarcely felt any more up-beat, but with a certain resignation declared: 'We have to prepare to meet a new order in this country.'[2] Unlike the churches in the rest of the United Kingdom, but like so many in central and eastern Europe, the Irish churches had to adjust to living under new regimes. Did those regimes take on the characteristics of confessional states – namely political structures where the tenets on one faith were enshrined in the law and the constitution?

When the Prime Minister of Northern Ireland declared in 1934 that 'we are a Protestant Parliament and a Protestant State', he was not indulging in triumphalist rhetoric. Indeed the Prime Minister who had come to power in Southern Ireland the year before – Eamon de Valera –

cherished a far more exalted vision of a confessional state, a vision that would be given some reality in the course of the 1930s and after. The Protestant churches in the south represented a tiny minority of the population, and therefore posed a minimal potential threat to any new political order there. But in the north they were representative of a majority which was faced with a large and disaffected minority. They used that position of influence – not always successfully – as the new government of Northern Ireland began to legislate for the particular problems of the Province in the early 1920s. Just a fortnight before the King opened the first Northern Ireland Parliament, in June 1921, the Presbyterian General Assembly considered 'A Programme of Social Reform' which 'in the interests of the kingdom of God, ought to be attempted, and may be achieved, within the next ten years'.[3] Industrial, social, moral and benevolent reforms were proposed, but education was first on the list. There was nothing sectarian about their proposed educational reforms. Seeking 'the fullest possible development of every child', they advocated the raising of the school leaving age to 16, provision of adequate educational and recreational facilities, abolition of child labour and strengthening of the School Attendance Act. The Assembly unanimously approved the proposals, on the motion of Professor Robert Corkey, whose brother William was to play a leading role in subsequent events.

A few months later the newly appointed Minister of Education, Lord Londonderry, began to assemble a team which would make recommendations regarding primary schooling in the Province, 'the first major task of reconstruction undertaken by the government'.[4] An unlikely reformer, a Tory grandee, cousin of Winston Churchill, educated at Eton and Sandhurst, he had already held ministerial office at Westminster, and now demonstrated his passionate devotion to the union by transferring his centre of activity to the new Belfast parliament. The committee he appointed was chaired by Robert Lynn, but the Roman Catholic Church declined to nominate representatives, a refusal that has been called 'the single most important determinant of the educational history of Northern Ireland from 1920 to the present day'.[5] The Presbyterian advice that the Roman Catholics should 'take their full share in moulding an educational system' went unheeded.[6] The one Roman Catholic member of the committee, T. Bonaparte Wyse, former commissioner of the Board of Education in Dublin and a kinsman of the Wyse who had fathered the national education system in the 1830s, was in no way representative of the views of the Roman hierarchy or of northern Catholics. So the committee was overwhelmingly Protestant;

it met over the subsequent nine months, against a backdrop of sectarian and republican violence, and of Roman Catholic non-recognition of the northern education system. Its interim report, published in June 1922 and broadly welcomed by the Protestant churches, envisaged the setting up of regional education authorities, which would disburse funds in proportion to the amount of control they exercised over schools. All recognised schools would have the salaries of teachers paid by the Ministry, and further aid depended upon the degree of control exercised by the new education authorities. Class I schools – whether built by the new local authorities or transferred by their existing managers – would be funded entirely, and be in effect 'state schools'. Class II schools were also called 'four and two' schools, with their management committees consisting of four members nominated by the manager (say, a church) and two by the Ministry of Education. They would receive, in addition to teachers' salaries, a grant for some 50% of their heating and cleaning costs. The Class III schools were like 'private schools' in England, though with their teachers' salaries fully subsidised. This new educational system, as envisaged by Lynn, and along with the other recommendations regarding attendance, meals and medical inspections, was as fundamental a change as the 1918 Fisher Act had been in England. But the contentious part of the Lynn report was that dealing with religious education, and in particular with regard to the Class I schools. The Bishop of Down, Grierson, expressed the majority view of the Committee (of which he was a member) when he said that 'religion should be taught in our schools' and that 'merely secular education does a cruel wrong to a child'.[7] No Roman Catholic bishop would have dissented from that view. The problem was: how should religious instruction to be given in 'state' schools? The committee's premise was that the old National School formula of 'combined moral and literary and separate religious education' should be maintained – when in fact it must have been clear to them that for the previous 90 years this had meant throughout Ireland a system of denominational schools, whatever the intention of its founders. The report envisaged 'a programme of religious instruction for their own children' to be formulated by the churches, though it was not clear whether this could or should be delivered within compulsory hours of attendance, and whether or not it should have specific denominational content. Nonetheless, the report was welcomed: a Presbyterian response considered 'its proposals sound and progressive. The one thing needful at present is a strong, enlightened public opinion that will claim their embodiment without delay in an Act of Parliament.'[8] They would have been less enthusiastic had they known what way the Minister of Education was proposing to handle the question of religious instruction. He was committed, as he told the cabinet when he presented his draft bill in December, to making 'public education purely secular'.[9] This would ensure that the act would not violate the constitution – whereby the Northern Ireland government was forbidden from endowing religion – but would allow the churches to make use of the school facilities to give instruction to pupils. When Robert Lynn saw the draft bill in February 1923, he wrote angrily to Londonderry, complaining about the religious education clauses and warning him that 'you are not prepared for the inevitable outburst of indignation from the people whom you will have so cruelly deceived'.[10] The indignation was not immediately apparent. The *Witness*'s editorial on the bill in March noted that it was 'favourably received by the House',[11] and in April during its second reading declared that Lord Londonderry's speeches had gone far 'to remove misapprehensions', and that 'many unfairly criticised the bill as if it were a scheme of purely secular education'.[12] The Revd William Corkey, soon to emerge as a leading critic of the government, decades later wrote his *Episode in the History of Protestant Ulster 1923–1947* as a vigorous defence of the line taken by the Protestant churches both in the inter-war period and in the immediate post-war years, though he glosses over the general acceptance of the bill in these early stages. It became law the day before the General Assembly met, and was broadly welcomed, though with some concern expressed regarding the appointment of teachers, and the financial provisions for the Class II schools.

It did not take long, however, for a head of steam to build up in opposition to the act. In July 1923 Presbyterian and Anglican representatives met to demand changes in the 'provisions made for the appointment of teachers and for the management of schools', recommending that meanwhile 'managers should not transfer their schools'.[13] Their approach to the Prime Minister, which was supported by the Methodists as well, drew no response, though privately Craig was saying that the Education ministry should look into the possibility of an amending act.[14] When no such amendment was forthcoming, the churches agreed in December 1924 to form a United Education Committee (UEC), steered by the convenors of education in the main churches – the Revds William Corkey (Presbyterian), J. Quinn (Church of Ireland) and W.H. Smyth (Methodist). Crucially, when the Committee met in February 1925, representatives of the Orange Order and prominent Unionists also attended. A mass meeting was arranged for March, and in preparation for it a vivid leaflet entitled 'Protestants

Awake' was produced. With an election in the offing, this would 'make the opposition quite clear to all those who would have the right to vote'. It warned of the threat to Protestant education in the province, with 'the door ... thrown open for a Bolshevist or an Atheist or a Roman Catholic to become a teacher in a Protestant school'.[15] The next day, the Prime Minister himself – in the absence of his Minister of Education, who was in London – met with representatives of the UEC and of the Orange Order, and then 'surrendered with almost dazzling swiftness'.[16] Within a week, on 13 March 1925, the Amending Act had been given the Royal Assent; and just over a month later, Craig won a resounding victory in the general election. The amending act, meanwhile, was open to varied interpretations, as became clear when Londonderry tried to explain it in parliament. After further negotiations a 'concordat' was agreed, which in effect set in concrete the state education system as implicitly a Protestant one. Local Education authorities were to require – in Class I and Class II schools – a 'programme of simple Bible instruction', and teachers were required to give such instruction as part of their duties. These requirements were anathema to the Roman Catholic church, and ensured that none of its schools would move from being Class III schools. It did however ensure that Protestant-controlled schools could with a clear conscience be transferred by their managers into the state system, with the greater funding which that entailed. It looked like a great victory for the Protestant churches. One historian has noted that 'no band of Catholic priests in the former united Ireland had engaged in politics with the energy and efficacy of the Protestant clerics who led the United Education Committee of the Protestant Churches'.[17] Another has pointed out that the public elementary system 'had as a result become denominational in all but name, at least for the Protestant community'.[18]

In fact, the debate was not yet over. In 1926 Lord Londonderry retired, not only from the Ministry of Education, but from the wider political scene in Northern Ireland. He was replaced, both in the Ministry and as Leader of the Senate, by another aristocrat, Lord Charlemont, whose sense of humour and deeper local connections made him perhaps a more approachable character, despite his being a Wykehamist. Changed too was the position of Roman Catholics, whose representatives now sat in parliament, and whose bishops were beginning to contemplate how they – like their Protestant counterparts – might gain some concessions from government. The issue that provoked the renewed debate was the refusal of the Armagh Education Committee to enshrine within transfer deeds a clause securing Bible instruction in the transferred schools in perpetuity. To this problem was linked the question of church representatives both on regional education committees and on the committee of management of Stranmillis Training College. The UEC swung into action again, and in July 1928, supported by Orange Order representatives, sought from Lord Charlemont an assurance that amendments to these ends would be introduced.[19] A joint delegation met with Craigavon (as Sir James Craig had become) and Charlemont in February 1929. The Prime Minister could foresee a re-run of the events of 1923–5, and again there was a general election in the offing. He temporised by suggesting that a small sub-committee might like to negotiate with the Ministry of Education, with a view to an amending act. The teachers unions were not at all keen on clergy representation – as of right – on education committees, and nor, it transpired, was the Orange Order. Within the cabinet, it was agreed that there was a 'need for a guarantee of security that Protestant transferred schools shall not in any circumstances get under Roman Catholic control'.[20] Such safeguards got a mention in the government's election manifesto, and the Education Act of June 1930 gave the Protestant churches most of what they had demanded. Representation (in practice, if not in law, clerical) was ensured on management committees of transferred schools, and those committees gained powers of appointment which in effect ensured that they could pick only Protestant teachers. Those teachers were now required to give undenominational Bible instruction, though pupils were not obliged to attend it. As a sop to the Roman Catholics, more public monies were to be available for voluntary schools. There was general approval of the new act which, as the Bishop of Clogher said, signalled 'the intention of the Government to guard against the secularisation of elementary schools'.[21] The General Assembly expressed its satisfaction that the new legislation would 'remedy the defects which have become apparent in the workings of the Acts' of 1923 and 1925.[22] But the question of clerical representation in the management of Stranmillis College had not yet been settled.

After partition, there was no facility for training primary school teachers in Northern Ireland, since both the colleges which Protestant trainees attended were in Dublin. The southern government decided to slim down its teacher training facilities, and while the Church of Ireland Kildare Place College would continue, the non-denominational (but effectively Presbyterian) Marlborough Street College was to close. During its final year, it was training 373 men and women, of whom a third were Presbyterians. The Belfast government arranged for the training of northern teachers at Kildare Place for a period of two years, while alternative arrangements were considered. The General Assembly in June 1922 called for a local teacher training facility, possibly in connection with Magee College, Londonderry.[23] However the government was moving swiftly, and by October 1922 had arranged for temporary teacher training facilities (administered by Queen's University and the Belfast College of Technology) which became Stranmillis College in 1929. The clergy of the UEC had insisted that there should be clerical representatives on its management committee. Lord Charlemont had agreed, but his cabinet colleagues did not, pointing out that Stranmillis was a government founded and funded institution. Rather than simply go back on the commitment he had made, the minister dissolved the Stranmillis management committee and brought the college directly under the control of his ministry in April 1931. Two months later the General Assembly debated primary education. It thanked the government for the new Education Act, but also stated that it was 'essential that such arrangements should be made for the training of Protestant teachers as shall command the confidence of the Protestant Churches inasmuch as such arrangements have been made for the training of Roman Catholic teachers as commands the confidence of the Roman Catholic Church'.[24] It was a familiar theme that had been repeated for a century: Presbyterians were in favour of non-sectarian education, but if the government acceded to the demands of the Catholic Church, it must in equity do the same for the Protestant churches. Sir Robert Lynn, at the Orange celebrations that year, claimed that 'Lord Charlemont had brushed aside the Protestant leaders with courteous contempt'.[25] Later that month, the government came up with a response. The Protestant churches were invited to nominate three clergy who would serve on a departmental committee of management for the College, but attending only when religious or moral issues were under consideration. A high powered meeting of the UEC convened in Belfast on 12 August. The Bishop of Clogher presided, and present were the Bishop of Derry, the Methodist President, the Principal of Assembly's and others. A letter was dispatched to the Prime Minister demanding further details.[26] Craigavon's response was tart. In a letter to Corkey on 21 August he maintained that 'Stranmillis is the property of the whole people. No person or association of persons has any right to special representation in the control of that institution.' His original offer still stood, which, as he pointed out, did 'not give the representatives of the churches any share in the general administration of the college'.[27] During the following months, as Corkey recorded, 'the *impasse* continued', though he does not record the split in the Protestant ranks. Archbishop D'Arcy, a close friend of the Prime Minister, was anxious to end the controversy, and declared himself satisfied. Others, especially the members of the UEC, kept up the pressure. As had happened before, they were aided by events outside their control. At the end of 1931, Cardinal McCrory created a storm of controversy by declaring 'that the Protestant Church in Ireland – and the same is true of the Protestant Church anywhere else – is not only not the rightful representative of the early Irish church, but it is not even a part of the Church of Christ. That is my proposition.'[28] Protestants were enraged, and further discomfited by two notable events in 1932, the first premiership of de Valera, who took office in March, and the Eucharistic Congress which was held in Dublin in June. In these circumstances it is not surprising that Craigavon wanted to see the Stranmillis agitation closed, and again took the initiative in settling it himself, in the absence of his Minister of Education. There was a need for 'strengthening the government's hands' and 'closing up the ranks of the Protestant and Loyalist community'.[29] To that end, Craigavon gave the UEC what it had demanded. As Corkey reported to the General Assembly in June, the three clerical representatives on the Stranmillis management committee would have 'full status, the same as the other members', noting that the settlement was 'entered into with the Prime Minister' and 'was approved by Lord Charlemont' and signed by the three representatives of the UEC, and by Sir Joseph Davidson on behalf of the Loyal Orange Institution.[30] Corkey spelt out the sectarian reality: 'We cannot afford to be divided in Northern Ireland and if the Government of Northern Ireland is to stand and be strong it must have the goodwill of the Protestant Churches.'[31]

That goodwill was strained again over education during World War II. In the dark days of 1940, 'Christian civilisation' itself seemed to be under threat. Archbishop Gregg opened the General Synod in May 1940 with a presidential address focused not so much on the war (he was, after all, speaking in a neutral country) but on education. He drew attention to the inculcation of pagan values in the youth of Germany: 'We need to be as good Christian citizens as the Germans are good Nazis.' Noting that 'there has been a tendency to concentrate attention unduly on the primary schools and to forget the secondary schools', he called upon the Synod, and parents north and south, 'to take every care that the programme of secondary education should include an irreducible minimum of religious instruction'.[32] A few weeks later, in the NI House of Commons, Professor Robert Corkey spoke of the failure of secondary schools 'to provide education on Christian morals which lie at the base of our democracy, backed up by simple and credible notions about God and the world in which we live'.[33] This was to become the new battlefield between the churches and the northern government during and after the war, with Corkey as one of the protagonists. The focus on secondary education was understandable. There had been growing concern about this sector of education both in England and in Ireland in the 1930s, and although the northern government in 1938 had planned to raise the school leaving age to 15, the war saw that plan abandoned. Secondary schooling in the north was structurally very different from the primary school system. Virtually all of the 70 secondary schools in the province were almost completely outside Ministry of Education control, 47 of them catering for the Protestant community. They might be subject to 'reconstruction' after the war, but meanwhile the feeling was abroad that they should be providing more adequate religious instruction. The General Assembly was told in 1941 by its secondary education committee that 'if civilization is to emerge from this generation strengthened and purified the place of religion in education must be made secure and that such religious teaching must be noteworthy for its sincere application of spiritual realities and sound scholarship'.[34] Thinking in England was moving along similar lines. The leaders of the churches there had written a letter to *The Times* in December 1940 urging educational reform, and while the General Assembly was debating in 1941, the new President of the Board of Education, R.A. Butler, was publishing a 'Green Book' laying out the ideas which would soon be embodied in the 1944 Education Act.[35] If Northern Ireland was to maintain a step-by-step policy with England, then the government must tackle the problem there too.

In May 1943 the new Prime Minister, Sir Basil Brooke, invited Professor Robert Corkey to become Minister of Education; he had represented Queen's University in parliament since 1929, winning his seat as a strong supporter of his brother's campaign for the amending of the Education Act. After 'considerable hesitation' he accepted the post, though there was some vocal opposition within the church when the General Assembly met a few weeks later.[36] Given the preparations in England for major educational reform, discussions got under way in the Ministry in Belfast for a similar measure in Ulster. Corkey was anxious to incorporate clauses dealing with 'Bible instruction and the appointment of teachers qualified to give it', and made clear to his ministry officials 'his determination to give all the children of Ulster in Primary and Secondary schools the benefit of Mr Butler's policy with regard to secular and religious instruction'.[37] For that reason, as he and his brother claimed, he was fired by the Prime Minister in February 1944, less than nine months after taking office. The official reason was that

Corkey had failed to give enough time to the job. Four months later the General Assembly passed a resolution expressing 'its appreciation of the great zeal shown by the Revd Professor Corkey in the discharge of his duties' and 'its satisfaction at the firm stand he took on the matter of religious education'. Their 'indignation at the treatment he received' was shared even by those who had the previous year queried the wisdom of a cleric taking up such a post.[38] The support for him was given even higher profile a year later when he was elected Moderator. The Methodists declared that 'the premier does not seem to have handled the situation in the wisest way', but concluded that it was hardly a church–state crisis.[39]

Lieutenant-Colonel S.H. Hall-Thompson, Corkey's replacement as Minister of Education, announced his proposals for educational reform in December 1944. In a major restructuring, akin to that which had already taken place in England, and in effect repealing the acts of 1925 and 1930, primary education would end with an 11-plus examination, and thereafter all children would proceed to 'intermediate' or 'grammar' schools, and stay at least until 15. Grants to voluntary schools for new buildings, maintenance, heating, lighting and cleaning would be raised from 50% to 65%. All schools would receive from reconstituted county and city education authorities free services in terms of school meals, books and medical treatment. There would be an act of collective daily worship, and religious instruction, in all schools. Teachers however would be protected, by a conscience clause, from being required to give such instruction. There were issues here which educationalists might debate, regarding the fairness of segregating children at 11-plus for instance, but predictably it was the religious issue which was to dominate discussion over the subsequent two years. And predictably too, the opponents of the changes found their leverage increased in view of the impending general election. A meeting with government was held in June 1945 from which the clerical and Orange Order representatives emerged content that a 'settlement agreeable to all' had been arrived at.[40] But by the end of the year, it was clear that the government intended to stick to its original proposals (especially as regards the wording of the conscience clause for teachers), and a year of 'sectarian bickering' followed. But the churches themselves were not united. When the General Synod met in May 1946, it welcomed the government's proposals, while opposing the proposal 'to establish Junior Secondary Schools in rural areas, believing that a better system of education might be evolved', and promising 'the most strenuous resistance to any attempt to alter the provisions for religious education as contained in the 1930

Education Act'.[41] Archbishop Gregg attended very few of the meetings between church representatives and the government, his biographer attributing 'his notorious absence' partly at least to his 'intense dislike of being, or being seen, in the company of Presbyterians or of nonconformists [*sic*]'.[42] The Presbyterians were divided. William Corkey led the opposition to the government's proposals, but the Revd John Waddell (Moderator in 1937) thought the fears being voiced were 'completely illusory', and publicly said so.[43] When the General Assembly finally welcomed the new Education Act, though with reservations, Waddell's attempt to expunge the clauses critical of government was defeated, but only by 172 votes to 105.[44] The Methodist Church welcomed the bill, supporting it fully when it became law.[45] The first of a series of rallies to protest against the government's proposals, held at Belfast's Wellington Hall in November 1946, revealed that the opposition included not only those who felt the government had reneged on its pre-election promises, but also those who had genuine educational and practical points of concern, as well as those who were simply opposed to any 'concessions' which would see greater financial support for Roman Catholic schools. Much of the opposition to the bill came from laymen like the rather maverick MP Harry Midgley (later himself to be Minister of Education) and the National Union of Protestants activist Norman Porter. The Orange Order did not mobilise in the way it had done in the past. The bill became law in March 1947. 'For the first time since 1925', writes Donal Akenson, 'the government of Northern Ireland had not altered its course in the face of an agitation by the more volatile Protestant elements'.[46] There was no further period of agitation. For the Minister of Education, however, it was not all over. Like his predecessor, he was sacked, but when that happened, it was not a result of church pressure.[47]

Had this 'episode in the history of Protestant Ulster' proved that Northern Ireland was a confessional state? The initial attempt to create a secular system of public education had foundered, as had the attempt via National Schools a century before, on the reluctance of both Protestant and Catholic churches to support such a system. The Protestants – and especially the Protestant clergy – had certainly proved that they had muscle, reflecting as they did the majority opinion among the Protestant majority. But their influence was limited, by government resistance to 'clerical' pressure, by the Government of Ireland Act prohibition on the endowing of religion, and by the clergy's dependence on the support of the Orange Order, which was not always forthcoming. The influence of the Protestant churches in other spheres gives a similarly ambivalent picture, which makes it difficult to uphold a coherent view of Northern Ireland as a confessing state.

During the debates on the education bill, Lord Glentoran had said: 'The trouble about us here in Ulster is that we get excited about education and drink.'[48] In the Presbyterian 'Programme of Social reform' of 1921, education had loomed large, while the drink question was merely one issue among many, under the heading 'Social and Moral Reform'. It proposed 'a measure of Local Veto, total Sunday Closing, and the abolition of all spirit-grocers' licences'.[49] That was a modest enough set of aims. The temperance movement was more ambitious, and retained much popular support. W.P. Nicholson's evangelistic campaigns helped to raise popular awareness, especially among working men, and astonishing tales were told and re-told about the transformed lives of men who had forsaken the demon drink. The Irish Temperance League placed great faith in the new government: 'on all questions affecting the moral and social well-being of the people over whom they ruled', Sir Robert Anderson told a meeting in 1923. It was 'out to see that the best laws were passed that could be enacted, in order to make it easy for people to do right and difficult for them to do wrong'.[50] The goal, which he optimistically thought could be reached, was that 'the liquor traffic should be cleared bag and baggage from the Province they so dearly loved'. A bill in the Commons in the spring of 1923 was introduced by the Prime Minister, who had faced strong opposition from temperance advocates when he stood for a Westminster seat in 1906; he was the son of a self-made whiskey millionaire, though he himself had no part in the drink industry, beyond inheriting a fortune from his father in 1900. Now Craig chose to pass a bill which would reflect much popular feeling in Northern Ireland, though it was by no means a 'bag and baggage' measure. As the Presbyterian newspaper said: 'There is nothing final in human things, and the present Bill is not the final word. It is, however, a good word, thank God!'[51] The churches expressed satisfaction at the abolition of spirit grocers' licences, long seen as 'an irredeemable public nuisance, and a source of temptation and danger to women customers'.[52] Sunday closure was introduced, as were restrictions on so called 'bona fide' travellers' ability to get drink in a hotel. The Bill would 'be greatly appreciated by the vast majority of Ulster people as it is in line with their cherished convictions', as the Chairman of the Presbyterian Committee on Temperance told the Prime Minister.[53] Its success from the churches' point of view may be gauged by the fierceness of the resistance put up by the brewers and licensees, mobilised in the Anti-Prohibition Council and the Ulster Reform Association. But the bill did not go far enough for some, and the demand for local option in particular kept rolling, supported by the General Assembly as 'the only final solution to the drink evil'.[54] A campaign was launched at a 'Six County Conference' in October 1926, with the aim of gaining local option by 1929, the year regarded as the centenary of the temperance movement in Ulster.[55] The Prime Minster made it clear that no further legislation could be expected from the government on the issue, which prompted the temperance advocate the Revd R.J. Patterson to ask a question which was to be repeated often in Ulster in the future: 'Are we all to be mum at every temperance meeting. And say not a word about legislation lest we should upset the Ulster equilibrium? Must we get permission of the cabinet before we dare to say we are going to demand further legislation? Must our loyalty to the Northern Ireland Government be manifested by our proving disloyal to our country?'[56] It was a good question. The answer has often been framed in terms of Northern Ireland as a sectarian state; it might also be seen in terms of the very small and parochial society that was Northern Ireland. In this case, the churches themselves were not united, the Anglicans being the most permissive on the issue. Many of them would have agreed with Bishop Hensley Henson who condemned 'the policy of Manichaean intolerance' which total abstainers and local option supporters pursued.[57] This was in the tradition of Anglican permissiveness, represented by the Irish Bishop of Peterborough, Magee, who famously declared to the House of Lords in 1872 that 'I should say it would be better that England should be free, than that England should be compulsorily sober', before dashing off to preside at the annual meeting of the Church of England Temperance Society.[58] When the Down Diocesan Synod followed in that tradition by refusing to back local option in 1927, the Presbyterians attacked the Church of Ireland in language reminiscent of the old days of Ascendancy-bating: 'Moderatism and flunkeyism in social life have been her bane. She has always been lukewarm in the furtherance of any Christly cause, and her advocacy of temperance has never been whole-hearted.'[59] The heat generated by the local option issue was also evident at the General Assembly debates in 1928 and 1929, but it had become clear to the Church leaders that local option was not within the realm of practical politics. The Primate, the Moderator and even the Methodist President declared themselves prepared to be guided by the Prime Minister on the issue, and asked their followers in 1929 to 'refrain from any action that would tend to divide the ranks of the supporters of the present Government'.[60] 'All the real facts are against local option', Bishop D'Arcy declared.[61] Had this whole issue been in any sense sectarian? Perhaps in one particular way it had. It reinforced what was so often seen by outsiders as the 'suffocating Puritanism' of Belfast and Ulster.[62] Sabbatarianism was reinforced by the Licensing Act, and the sectarian divide was thus widened, because as Mary Kenny notes, 'Sabbatarianism and temperance were prime concerns of Irish Protestants at this time, whereas for Catholics, bad books and immodest frocks took precedence.'[63] The character and style of Protestant Ulster was evidenced in a continuing campaign to keep Sunday pure. Even in holiday resorts, Sunday concerts were banned, as they were at Portstewart in 1930, despite the complaints from the manager of a pierrot troupe about 'the dullness of Sundays', and his claim that holidays were 'for smiles, not boredom and faddism'.[64] Along the coast at Portrush, 30 years later, the Council decided to permit Sunday variety concerts, but local ministers complained and got up a petition, and 'thus the desecration was prevented'.[65]

While Northern Ireland may only to a limited degree have been a confessing state, it was certainly a society riven by sectarianism. That was evident in the inter-communal strife which attended its birth. It was evident also in the events surrounding the Eucharistic Congress of 1932. International gatherings of that name had been held by the Roman Catholic Church since the late nineteenth century, to promote devotion to the Blessed Sacrament. The London Congress in 1908 stimulated a wave of anti-Catholicism when 'religion had proved to be a decisive and divisive political issue in a society which prided itself on its secularism and relative tolerance'.[66] A quarter of a century later in Ireland, it was likely to have similar results, even if it was taking place in the staunchly Catholic south. The northern government was faced with a dilemma when it received invitations for its senators, MPs and others to attend the proceedings; rather than face the embarrassment of saying 'no', the answer was 'to completely ignore the invitation'.[67] The event could certainly not be ignored in the south, as a huge wave of devotion swept over the country. It affected even Protestants. The Rector of Trimoleague's wife 'was with difficulty restrained from hanging a Papal flag out of the rectory window'.[68] Even in Fermanagh, Protestants 'showed signs of good-will towards their Catholic neighbours, helping them in some instances with the decorations'.[69] The Anglican Bishop Day of Ossory thought that 'even the strongest Protestants must have been stirred by the thought of that vast multitude, gathered from every town, village and countryside of Ireland for the worship of Christ the King'.[70] Some could not contain a laugh or two: there was a report of a banner in O'Connell Street proclaiming 'God bless the Holy Trinity', and even the Catholic convert G.K. Chesterton reported hearing a woman on a bus saying 'Well if it rains now, He'll have brought it on Himself.'[71] Sadly, there were ugly scenes too, with northern Catholics attacked as they boarded trains to go to Dublin. The most vicious attack was on a procession of pilgrims returning from the Congress when they offloaded at Larne. The Governor expressed his disquiet. The Catholic Bishop of Down and Connor complained bitterly, though the response (from an Assistant Secretary at Stormont) was that if the police had known that the clergy intended to process wearing vestments, the procession would have been banned, as 'this was calculated under the circumstances to lead to a breach of the peace, and is in addition contrary to law, as Your Lordship is no doubt aware'.[72] In the south meanwhile, de Valera took a leading part in the Congress, and 'derived untold benefit from the principle of piety by association'.[73] He used the occasion to humiliate the Governor-General, James McNeill, and to set in train the events which led to the abolition of the office which was one of the few vestiges left of Ireland's links to the Crown. For northern Protestants, and even for southern ones, all this could only reinforced their 'home rule means Rome rule' prejudices.

The resurgence of Catholicism, and the threat of a more aggressive government in the south, combined with economic depression to make the 1930s a time of great volatility in the Province. Economic hardship was grave even by United Kingdom standards. Unemployment in the shipbuilding sector reached an all-time high of 64.5% in December 1932. Northern Ireland had, as Patrick Buckland has said, 'the most divided and disadvantaged people in the United Kingdom'.[74] It was fertile ground for the recrudescence of sectarian violence. The Ulster Protestant League (UPL), founded in 1931, and its short-lived and even more extreme offshoot, the Ulster Protestant Society (UPS), founded in 1932, contributed to the sectarian volatility of these years.[75] However deprivation affected both Protestants and Catholics, and the banning by the government of a hunger march of the unemployed in October 1932, led to a rare example of a non-sectarian disturbance, in which two workers were shot dead, and many wounded and injured. But the sectarian pattern soon re-established itself. The UPL staged meetings in May 1934 to protest against a Catholic Truth Society Festival being held in Belfast, and one of the UPL speakers, the Presbyterian Revd Samuel Hanna, was convicted of incitement. The other cleric who figured prominently on the UPL platforms (and briefly with the UPS) was the Revd John Glass, Methodist minister in Donegall Road, noted for his 'strong evangelical preaching' and his desire to 'relate the Gospel message to social conditions'.[76] It was no doubt social conditions that helped foment unrest in Belfast's working-class areas; the UPL both blamed the northern government, and railed against the revanchism of the de Valera regime in the south. When the Twelfth of July march was banned in 1935, and then unbanned, the result was an outbreak of sectarian violence which left thirteen people dead, and hundreds injured and driven from their homes. The Roman and Anglican bishops of Down and Connor, Mageean and MacNeice, co-operated in appealing for calm among their respective flocks, though they saw the problem very differently; three years later Mageean was appealing to de Valera to end partition – 'an evil which only its removal can remedy'.[77] The Presbyterian General Assembly, meeting almost a year after the riots, made no mention of them, and while pointing out that 'the increase in juvenile crime in Belfast is most alarming', attributed that to 'gangster films as well as lack of parental control'.[78] The Report of the Council of Churches came no closer to analysing the problem, focusing as it did on the 'affliction and persecution' of Russian Christians, on 'Christian refugees from Germany', and on world peace. The nearest it came to dealing with Northern Ireland's own problem was to report that a Belfast Council of Churches had been set up, 'to promote good-will in the community' and to 'ensure that the fundamental rights of civil and religious liberty and personal safety shall be assured to all citizens'.[79]

World War II for most Protestants, certainly for northerners, was a morally justified crusade against evil forces, a war to preserve their 'civil and religious liberty'. J.F. MacNeice had preached with some prescience in 1933 on the failure of politicians and churchmen in Germany. 'Lutheranism in Germany', he had said, 'was too individualistic and witnessed very feebly and ineffectively for the application to all life, social and national, of the principles of the Christian religion.'[80] That was a charge which was to be made against Irish Protestant churches in the years ahead. But they could not be accused of flirting with fascism, even if an organisation like the Ulster Protestant League 'has usually been mistakenly labelled as fascist'.[81] The gulf between Protestantism and Roman Catholicism in Ireland was widened by the different approaches to world politics in these years. The idea of a corporatist state, with its echoes in Dolfuss' Austria and Mussolini's Italy, and Pope Pius XI's 1931 encyclical *Quadrigesimo Anno*, was taken up in Southern Ireland with some enthusiasm, Fathers Edward Cahill and Denis Fahey being among the chief protagonists.[82] Their ideas were corporatist, anti-Communist and anti-Semitic.[83] Irish nationalism could relate to all this. The northern Nationalists made clear in 1938 that they would oppose conscription, as they had done in 1918, and they even compared their plight to that of the Sudeten Germans, and in 1940 were anxious 'to place the Catholic minority in the North under the protection of the Axis powers'.[84] That attitude, and the fact that Ireland was the one part of the Empire and Commonwealth which did not declare war on Nazi Germany, created not only a greater gulf between Catholics and Protestants, but also 'reinforced the psychological gap between Ulster unionism and Irish nationalism'.[85] Protestants viewed the war as a crusade for freedom, and an opportunity to express loyalty to the Crown. And when the USA entered the war, there was a chance to gain the goodwill of America, whose troops were stationed in Northern Ireland and who were never left unaware of Ulster's solidarity as opposed to the South's non-belligerence. At a different level of solidarity, Miss Thelma Smith of Belfast married Private Herbert Cooke of Ohio in College Square Presbyterian Church, and became the first of an eventual 1,800 GI brides.[86] The visit of Cardinal Spellman of New York to the US troops in Ulster posed something of a dilemma for the government in 1943. The cabinet debated whether they should take any notice of it: for the Prime Minister of Northern Ireland to greet publicly a Catholic cardinal would not go down well with his constituency; at the same time, American goodwill had to be milked. They opted for 'the course which would arouse least criticism', which was a private lunch in the Prime Minister's room at Stormont.[87] Another Roman Catholic caused embarrassment to the Unionists, when James Magennis became the only Ulsterman to win the Victoria Cross in the course of the conflict. He received scant recognition in the city of his birth, and it was not until the end of the century that a public memorial was unveiled to him in the City Hall grounds.[88]

The war ensured higher rates of employment and greater prosperity, but it also brought disruption, death and destruction. The disruption was soon evident. The Trinity College Mission in Belfast reported that its various organisations, 'except the hockey club' were continuing as usual, though the attendance at the men's Bible class had been halved, because men were working longer hours, and night shifts.[89] Evening services were soon hit by the blackout. Sunday schools were disrupted by the evacuations of school children. One minister considered that 'Sunday schools and Sunday evening worship really never recovered again.'[90] People prepared themselves for the bombs they knew would come, and one church used the Air Raid Precautions (ARP) to drive home a message:[91]

Acquaint thyself with God and be at peace!
Repent and be converted!
Prepare to meet thy God!

This last advice was all too appropriate because in April/May 1941, Belfast suffered some of the most concentrated and devastating air raids of the war. Over a thousand people were killed, and over half the city's housing was damaged or destroyed. In all, seven Presbyterian churches were destroyed and eleven were seriously damaged. Three Methodist churches and four Anglican ones were destroyed. The suffering of people, killed, bereaved, injured, or deprived of their homes, cannot be counted denominationally. There was a shared sense of suffering among Protestants and Catholics; or was it simply a shared panic?[92] Certainly bombs were no respecters of persons. Certainly there were men from the south fighting in the armed forces alongside northerners. There were fire engines from the south aiding Belfast during the blitz. Dublin even suffered an air raid. But all this did not lead to closer relations within the north, or between north and south. Three years after the war ended, Ireland had left the Commonwealth, and a year later, Westminster had guaranteed that Northern Ireland would not cease to be a part of the United Kingdom 'without the consent of the Parliament of Northern Ireland'. As after World War I, so after World War II, the Irish Churches entered a different world. 'We meet today under strange conditions, with the Republic of Ireland an established fact', said Primate Gregg at the General Synod in 1949, though he emphatically reaffirmed 'the essential oneness of the Church of Ireland ... "Hands across the border" must be the unfailing principle of our common Church life.'[93] Gregg himself had experienced the 'changes and chances of this fleeting world' at a much more personal level. His younger son, Claude, had died while a Royal Navy trainee back in 1928; in January 1944 he heard the news that his elder son, John, had died in the Pacific five months earlier; in August 1945 his wife died after a long illness. Whatever his granite exterior, Gregg was suffering loss and grief at the end of a war which had brought so much grief and loss to many millions of people. The historian may seek to record the great events; for most Irish Protestants, personal concerns and local issues were of more moment. In the bleak post-war years, the freezing winter of 1947 was of more consequence than the Cold War.

The churches in a Roman Catholic state

The south had been, Emmett Larkin argues, a confessional state *de facto* since the days of Parnell's enrolment of the Roman Catholic Church in his home rule campaign, but 'it was left to de Valera to make that state as formally confessional as it had been informally since 1886'.[94]

Corporatist ideas of the state were an inter-war reinterpretation of traditional Catholic teaching, and the new Irish state comfortably accommodated such thinking. For Protestants living there, it only increased the difficulty of adjusting to being ruled by a Dublin parliament, which most of them had never wanted in the first place. It was 'all very well to say that loyalty can be transferred', said the *Church of Ireland Gazette*, but loyalty 'is an affair of the heart and it is not possible to force the heart to follow the hand'.[95] Reluctantly they accepted the new regime; but their hearts were not in it, and its increasingly 'Catholic' colour did little to persuade them differently.

The Irish language was to be a key area. The Gaelic League, founded in 1893 by Eoin MacNeill, an Ulster Catholic, and Douglas Hyde, son of a southern Anglican rector, played a significant role in the development of the concept of Irishness, and in the emergence both of Sinn Fein and of the cult of anti-Englishness. Although the Gaeltacht area was in the west of Ireland – a place of 'barefoot children, turf fires and unrelieved diet'[96] – the romantics and scholars who propounded the importance of the native tongue were often bourgeois Dubliners. In the early days of the League they were non-sectarian and apolitical. Two Protestant members, Sean O'Casey and Ernest Blythe, helped set up a service in Irish at St Kevin's, Dublin. The Revd J.O. Hannay became a high-profile Protestant member of the League, and managed to arrange for an early morning celebration of Holy Communion in Irish at St Patrick's Cathedral on St Patrick's Day, 1906, though the Dean, Dr Bernard, told Hannay that in permitting it he had 'been accused of condoning a profanation of the Blessed Sacrament'.[97] When the Presbyterians were approached by the League in 1905 with a view to their holding a St Patrick's Day service in Irish, their negative reply neatly encapsulated the attitude of many Protestants. 'The Presbytery believe that a principal cause of the decay of the Irish language in Ireland is the fact that it is not employed in the religious services of the people', they noted caustically, pointing out further that Presbyterians in the past had encouraged the use of Irish, but this no longer made any sense since 'Irish is no longer understood by the people with the exception of a very few.'[98] Like the Presbyterians, Anglicans had been interested in the use of Irish in the previous century, and it was proselytising motives that lay behind the establishment of a chair in Irish at TCD in 1838. But there had been clergy with a genuine interest in Irish language and lore, such as the Revd Maxwell Close and the Revd Euseby Cleaver, while the Revd Robert King of Ballymena published an Irish translation of the Book of Common Prayer in 1860. When a lectureship in Celtic was established at Queen's, Belfast in 1909, the first occupant was an Anglican cleric, the Revd F.W. O'Connell. The Irish Guild of the Church was an example of how the new mood, the 'Keltic Revival', affected Anglicans, though as we have seen, in the end it split over the political connotations of 'Irishness'. It launched *The Gaelic Churchman* in 1919 with the intention of countering 'English political idolatry' in the Church of Ireland, for – as a letter to the editor claimed in its first issue – if they did not, 'we may see substituted for her an English Colonial Church with German Lutheran characteristics'.[99] That kind of rhetoric alienated many ordinary Anglicans, and Gregg must have bridled at its attack on his address to the Diocesan Synod in 1921, described as demonstrating 'in a lamentable way the attitude which the Church of Ireland has adopted in considering itself alien and aloof from the majority of its fellow countrymen'.[100]

When the Free State authorities took over responsibility for education in 1922, it soon became clear that its educational reforms were 'focused almost exclusively on transforming the schools into key agents of the revival of Irish and Gaelic culture'.[101] In December 1922, Archbishop Gregg led a deputation in what one of the participants (the Revd H. Kingsmill Moore) described as 'a dramatic interview' with the Minister of Education, Eoin MacNeill – he who with Hyde had started the Gaelic League. Gregg 'opened with a vigorous and at times indignant exposure of the way in which our children were being compelled to study Irish books written for RC pupils'.[102] MacNeill was conciliatory and 'gently promised protection', so the Church of Ireland Board of Education was able in February 1923 to circularise all 'clergymen whose children are compelled to attend schools not under Protestant management', advising them that they should ask the managers of such schools to use 'phrasebooks such as that of Father O'Leary, "An easy Irish Phrase-Book", free from sectarian bias'.[103] The Revd A.A. Luce, Fellow of TCD, was still complaining three years later that compulsory Irish was 'a wrong to the religion of Protestants' because all 'the associations of compulsory Irish are Catholic' and there was 'little or no Protestant literature in the language'.[104] The Church was addressing the problem. In 1927 a preparatory college, Coláiste Moibhi, was set up at Glasnevin to supply the Church of Ireland Training College with trainee primary school teachers who were fluent in Irish. When in 1928 it was proposed to make Irish a compulsory pass subject in order to pass the Intermediate Certificate, there were protests in the Dail.[105] Gregg, and indeed the Provost of TCD, were inclined to accept the change, which led the *Irish Times* to fulminate that compulsory Irish was 'a denial of intellectual freedom' and 'a material menace to the Church of Ireland's youth'. Yet, it complained, 'the church's leaders are dumb'.[106] This was not entirely fair. Bishop Patton of Killaloe told his Diocesan Synod in July 1929 that 'a government which, whether by a jackboot or a thumbscrew policy, enforced such a policy was plainly a coercion government ... the compulsory teaching of Irish, as now prescribed, was a tyranny'.[107] That 'tyranny' was intensified when Fianna Fail came to power in 1933, its Minister of Education, Tom Derrig, 'noted for his zealous commitment to Gaelicisation'.[108] The Protestants found themselves so limited in the choice of textbooks which would be acceptable both to them and to the inspectors of schools that a competition was organised to find a suitable history textbook. It was won by Miss Dora Casserley, an enthusiastic member of the Irish Guild of the Church, whose book went into publication in 1941. It 'helped Protestant children to develop a pride in the contributions of some of their co-religionists to the national struggle', and by 1947 had sold 26,421 copies. In the following year, the Department of Education adopted it as a sanctioned book on its official list.[109] The problem persisted in that many prescribed textbooks were soaked in Catholic assumptions and replete with Catholic references. The Department of Education was not of a mind to help, and indeed suggested that the Protestants could have their own texts in Irish published – which was not an economic proposition.

On one issue the Department was thwarted. Derrig introduced a Schools Attendance Bill in 1942 to give him power to certify that the education received by a child was 'suitable even if they were sent to schools outside the state'. It was, de Valera admitted, an attempt to ensure that a gap was closed in law, whereby children might avoid having to learn Irish by being educated in Northern Ireland or Britain. Passed by both Houses of Parliament, it was referred to the Supreme Court who ruled it *ultra vires* in terms of article 42 of the Constitution, which acknowledged that 'the primary and natural educator of the child is the Family'.[110] Ironically the President was the Gaelic-speaking Anglican Douglas Hyde, and the Constitution which protected the Protestants was that drawn up by de Valera to enshrine his ideals of a Catholic and Gaelic Ireland.

Protestants were not entirely out of sympathy with some 'confessional' features of the Irish Free State. Censorship is a case in point. Films were subject to censorship from 1923. The Protestant churches were at least as vigorous in their condemnation of the cinema as the Catholic Church. At a great 'public indignation meeting' before the war, the proposed showing of a film about the life of Christ was attacked as being a sacrilegious money-making venture, which exposed clearly the dangers of lack of censorship.[111] 'There is no more insidious danger to public and private morals than the cinematograph', declared the *Church of Ireland Gazette* in 1925.[112] When J.R. Mott, the American evangelical leader, visited Belfast in 1930 he blamed his own country for producing films which 'undermined in a night what had taken the missions years to build up',[113] and in 1931 the General Assembly called on the northern government to introduce 'rigid censorship of Films, Books and Literature, calculated to have an evil effect on the morals of the rising generation'.[114] The following year the *Irish Churchman* expressed deep concern at the appearance of films with 'significant titles' like *Loves of an Actress, Beware of Married Men, Powder my Back*, and *A little Bit of Fluff*.[115] The Presbyterians lamented 'the fact that life's most sacred relationship is often sullied as a result of the suggestive nature of films thrown on the screen', congratulated the southern government 'on the strong stand they have taken on this matter', and called upon the northern parliament to institute 'stricter censorship'.[116] It took the Presbyterian Church a long time to become reconciled to the undoubted popularity of the cinema. In 1945 it acknowledged 'the power and influence of films', which 'should be used to the full in the cause of morality and religion'.[117] When the Methodists started showing films at Belfast's Grosvenor Hall in the 1950s, their 'cinema mission' attracted audiences of over a thousand, as well as pickets declaring the evil of such performances.[118] Even in 1977 there were vocal protests against the showing in Kilkeel of *The Sound of Music*, the Democratic Unionist Party denouncing it as 'full of Romish influences which Protestants abhor'.[119] Presumably it could not be attacked on the grounds of sexual impurity.

Sexual impurity, and the corruption of public morals, was the concern of the Committee of Enquiry on Evil Literature set up by the southern government in 1926. It received no evidence from the Protestant churches, though the Dublin Christian Citizenship Council was represented. Its spokesman, Dean Kennedy of Christ Church, was not too worried about objectionable publications as 'they did not seem to be read very extensively by members of the Protestant community', but he did touch on the issue of birth control. Here he had to admit that 'among Protestants there was a divergence of opinion'.[120] The Committee's recommendations resulted in the Censorship of Publications Bill, given its first reading in the Dail in July 1928. The Protestant churches did not object, though Sir John Keane, a leading member of the General Synod, passionately attacked it in the senate, and leading literary figures like George Bernard Shaw, W.B. Yeats, Oliver St

John Gogarty and George Russell all voiced opposition, to no avail. The bill became law, and the Censorship Board of five members began work in 1930; Professor Thrift of Trinity, who had served on the Committee of Enquiry, was its one Protestant member, and the northerner Leslie Montgomery (who wrote as Lynn C. Doyle) later served briefly. Opposition to censorship was not a case of Protestants versus the Catholic state. It came from liberals, literary figures and those who objected to what they saw as the stultifying narrowness of the Irish state. When Sean O'Faolain, Gaelic Leaguer and ex-IRA man, met de Valera again in 1933 after ten years, he could only reject his 'pietistic simplicity' and intellectual obscurantism, and all the things that had flowed from them: 'censorship, conservatism, the indolent glorification of all things Irish and the sentimental, low church Catholicism'.[121] The Protestant churches might feel uncomfortable in de Valera's Catholic Ireland as it was developing in the 1930s, but they found it hard to object to the specific steps taken to buttress it. From 1925 divorce was impossible in the Free State, but the Protestant churches were not (at this stage) going to object; nor were they going to come forward as champions of abortion or birth control. The sale of contraceptive devices was made illegal in 1935, despite an 'extensive demand for contraception among the urban middle classes',[122] but it was only in the post-war era that Protestant churchmen raised the issue as a matter of individual freedom. The Catholic clergy condemned dancing. The *Irish Churchman* agreed, admitting that while dancing could be an innocent and enjoyable pursuit, 'with the strange perversity of our kind, we turn it, only too often, into a curse'.[123] In 1935 the government, once again 'using legislation as an agent for moral reform',[124] made public dances illegal without license from the district courts.

Some issues caused Protestant Churchmen to speak out. Gambling was 'the greatest evil of the moment', the Presbyterians declared in 1928.[125] It 'had become the greatest social evil the Churches had to fight', the Rector of Christ Church, Londonderry, declared in 1930, it had 'become a greater menace than the drink evil', responsible for 'breaking up homes, and destroying the souls of men, women and children'.[126] The Free State legalised off-course betting in 1926, and hospital sweepstakes in 1930. 'If', Archbishop Gregg declared in 1929, 'a little more of the energy that was spent in Ireland in denouncing certain other dangers ... were devoted to attacking and curbing the vice of gambling, many Irish homes might benefit and many Irishmen be rescued from moral breakdown.'[127] Unable to shift the government on the issue, the Presbyterians at least ensured that their own house was in order. In 1931 they ruled that no bazaar or sale of work could take place on their premises unless the 'promoters have undertaken that no money shall be raised by means of balloting, raffling, or lottery tickets'.[128] The Methodist Church made similar provisions.[129]

The Catholic and Gaelic Ireland in which Protestants might feel uncomfortable had not taken shape in any constitutional form in the 1920s. Southern Protestants were fairly well represented in the Dail, with 13 or 14 members out of 153 throughout the 1920s. They were even better represented in the Senate, where in 1922 they numbered 24 out of a total of 60. Protestants could feel they had a voice proportionately greater than they might have expected, even if their general policy was to 'Lie low and say nothing.'[130] They cherished the enshrined right of appeal to the Privy Council, which has been described as 'the most obnoxious feature of the Constitution' as far as the Irish Government was concerned.[131] Irish delegates raised the matter at the Imperial Conference in 1926, which concluded that no changes should be made in any one member state without further discussion by all. Before the next Imperial Conference of 1930 the two Anglican Archbishops spoke out, acknowledging 'the fair and generous way in which we as a minority have been treated', but viewing with 'very real concern the avowed determination of the Government to deprive the citizens of the State of the right of appeal to the Privy Council'. They went on to speak of 'grave disappointment' and even 'alarm' at the possibility of the abolition of this 'most valuable protection to the minority which we represent'.[132] In the event the 1930 Conference introduced no changes, but the subsequent Statute of Westminster (1931), a milestone in the development of the British Empire into the Commonwealth, acknowledged that the parliament of the Irish Free State (and those of Australia, Canada, New Zealand, South Africa and Newfoundland) had full sovereignty. So when de Valera came to power in the following year, he set about removing the structures and symbolism which had meant much to the southern Protestants. James MacNeill, the Governor General, deliberately humiliated during the Eucharistic Congress, was forced out at the end of 1932 and replaced by Donal O'Buachalla, 'a retired Maynooth shopkeeper', who on a modest salary lived inconspicuously in a little Dublin house until the office was abolished four years later.[133] Early in 1933 the Oath of Allegiance was removed from the Constitution, and in November the Governor General's right to withhold assent from bills was abolished, as was the right of appeal to the Privy Council. In May 1936 the senate was abolished and in December – in the wake of the abdication crisis – the remaining references to the King were removed from the constitution.

The new Constitution now proposed by de Valera began with a preamble '[H]umbly acknowledging all our obligations to our Divine Lord, Jesus Christ, Who sustained our fathers through centuries of trial' and '[G]ratefully remembering their heroic and unremitting struggle to regain the rightful independence of our Nation.' That was language alien to most Protestants.[134] The actual articles of the Constitution also caused some concern. De Valera was anxious to pay due homage to the Catholic Church. There were those who wished to see the law of the Church embedded within the constitution, and disliked any reference to 'other churches'.[135] De Valera, a faithful son of the church, was also a politician. He visited Archbishop Gregg – the first meeting of two men who were often mistaken for each other – and subsequently credited Gregg with the idea that the new Constitution should 'give each Church the title which it had formally given to itself'.[136] He consulted also with the Presbyterian minister in Lucan, Dr J.A.H. Irwin. When in 1919 some Protestant clergy, and Sir Horace Plunkett, had toured America in the hope of counteracting the high profile which de Valera had achieved there, Irwin had gone to speak from de Valera's platforms 'in defence of the proposition that the struggle in Ireland was a national, not a religious, issue'.[137] Unpopular with his fellow churchmen on that occasion, he was now valued since his 'personal friendships with leaders in political and public life in Dublin enabled him to represent the Church there with success in many important occasions'.[138] De Valera also met with a deputation which included the Presbyterian moderator-elect, John Waddell, who was related by marriage to Irwin. The wording that appeared in article 44 steered a middle course between the demands of some Catholic churchmen, and the natural reservations of non-Catholics. It recognised 'the special position of the Holy Catholic Apostolic and Roman Church as the guardian of the Faith professed by the great majority of the citizens', and also recognised 'the Church of Ireland, the Presbyterian Church in Ireland, the Methodist Church in Ireland, the Religious Society of Friends in Ireland, as well as the Jewish Congregations and the other religious denominations existing in Ireland at the date of the coming into operation of this Constitution'. Grave offence to both sides was avoided, though what 'special position' meant was not entirely clear. It became clearer in other articles which attempted to flesh out de Valera's vision of a Catholic Ireland. Divorce or the remarriage of persons divorced elsewhere was forbidden. Other articles concerning the family and education were seen both as incorporating a

Catholic vision of the family, and as discriminating against women – a point made at the time, and not merely in later years when feminism was fashionable. Its claim to the whole of Ireland infuriated Ulster Protestants, even if it was tempered by a recognition that the *status quo* was rather different. When the Constitution was presented to the electorate in a referendum, held on the same day as the general election, it was approved by 685,000 to 527,000, though probably 'the overwhelming majority of Protestants voted heavily against'.[139] However slim the margin, Ireland was now effectively a republic, and de Valera was returned again as Prime Minister, or 'taoiseach' (leader) as the new constitution – betraying here its 1930s origins – called him. His party, Fianna Fail, had lost seats however, which is perhaps an indication that it was not only Protestants who looked askance at the new Constitution.

Protestants had not suffered ill-treatment or discrimination in the years following the violent establishment of the Free State back in the early 1920s. There were occasional incidents. In 1929 Mrs Trench, whose husband had been a JP, was forced out of her home in Co. Limerick by threats from a man who, when she returned home not wishing to pursue the matter, was given a suspended sentence. Kurt Bowen has noted that the real significance of the case is 'the typical unwillingness of Mrs Trench to draw attention to herself and her community'.[140] Or, as W.B. Stanford put it in 1944, the 'church of the governing classes had become the church of a politically discredited minority of one in twenty', so the 'general policy adopted was "Lie low and say nothing"'.[141] The sectarian violence in Belfast in 1935 caused a reaction in the Free State, with attacks on some Masonic halls and Protestant churches. De Valera 'expressed his regret and arranged for the government to make good the damage'.[142] There was an attempt at priestly pressure on the *Irish Times*, when it published articles on the Spanish Civil War which were deemed to be biased. Bert Smyllie, the assistant editor, showed the priest to the door and 'hardly before he was out of earshot', broke into the song:[143]

Oh, the Pope of Rome I do defy,
And every Papish union,
And as I live I hope I'll die
A loyal Presbyterian.

But now that the new Constitution was in place, would the pressure increase? Was Eire now 'a clerical state'? Was Canon Boylan right when he announced in 1937: 'The Catholic Church is more favourably placed in Ireland than it is in even the most Catholic countries of the continent.'[144] It began to look like it. Ireland's neutrality during World War II had the effect of isolating her from the mainstream of world developments. The post-war years saw higher rates of unemployment and emigration, but in the 1950s the Catholic Church seemed to be at the height of its power; 'religion was the bastion of security for a demoralised people'.[145] In the late 1950s, one secondary school leaver in every eight entered a seminary. The 'Mother-and-Child' controversy of 1950–1 seemed to demonstrate how powerful the church was in post-war Ireland, or, as one critic claimed, it 'revealed how meaningless the religious-liberty guarantees of the Constitution could be when the hierarchy chose to sidestep them'.[146] Dr Noel Browne, the somewhat maverick and prickly Minister of Health, proposed a bill giving free medical care for mothers and children on a non-means-tested basis. The opposition of the hierarchy, not to mention his own poor communication with his colleagues, led to Browne's capitulation and resignation. The *Irish Times* claimed it as evidence that the Catholic Church seemed to be 'the effective government of the county'.[147] The Protestant Churches avoided public comment. De Valera, who as leader of the opposition had 'played a waiting game of masterly inactivity'[148] during the controversy, seems to have gained some credibility among the Protestants, for at the election of 1951, they were said to have voted for his party, 'feeling that at least he could be trusted to stand up to the bishops'.[149]

There were in fact some issues which affected Protestants more directly. In 1949 some of the Knights of St Columbanus managed to gain control of the mainly Protestant Meath Hospital, by exploiting its rather lax membership rules. The government came to the rescue by passing a Meath Hospital Act which ensured that control would remain with its traditional management.[150] Two marriage cases also caused concern. In February 1945 a marriage between a Roman Catholic and a Presbyterian took place in the Presbyterian church in Drogheda, but in March 'the Irish Press carried a photograph showing the couple coming out of Trim [Roman Catholic] Cathedral'.[151] The matter was taken up by the Revds John Barkley and J. A. H. Irwin as being a breach of the Constitution, since it suggested that the Presbyterian marriage was invalid. In 1991 Barkley remembered that 'Mr de Valera assured the Presbytery, through me, of the validity of Presbyterian marriages. A satisfactory outcome.'[152] However writing in 1993 he reported that although he and Irwin had been assured, and had 'accepted in good faith', that 'all reference to the second marriage will be removed from all State records', this had not in fact happened. So 'when tested recently it was found possible to procure a Certificate for the 13 March date'.[153] The Tilson case in 1950 was rather more serious. Ernest Tilson, an Anglican, married a Roman Catholic and gave the required pledges that any children of the union would be brought up as Catholics. When the marriage broke up, the father took three of their four children to be brought up as Protestants, and the mother sought custody. Both the High Court and the Supreme Court ruled against the father, although the only Protestant Judge involved registered his dissent. It looked as if 'the courts had abandoned a rule of law which was impartial between denominations in favour of one which benefited the Roman Catholic Church'.[154] All this made for Protestant unease. So did the perceived power and influence of movements like the Knights of St Columbanus, and of Maria Duce. The former had shown its clout over the Meath Hospital. The Maria Duce Movement, founded by the right-wing Fr Denis Fahey, was at its most militant in these years, campaigning for the emending of article 44 of the Irish Constitution to bring it more into line with Catholic teaching. 'For a Catholic', declared the secretary of the movement in 1950, 'religion is a matter of dogmatic certitude. For him there is only one religion. In consequence all non-Catholic sects, as such, are false and evil.'[155] When in November 1950 the Pope promulgated the doctrine of the bodily Assumption of the Blessed Virgin Mary, 'Protestant unease was great.'[156] Primate Gregg wrote immediately to his clergy 'peremptorily forbidding any of them to reply to it'.[157] The Church of Ireland bishops then issued a Pastoral letter, as drafted by Gregg, to be read in all churches in December, refuting the new 'dogma', and noting that the Old Catholic churches, with whom they had relations of inter-communion, had been outspoken in their rejection of it. For Old Catholics as for Anglicans it was a question of authority: 'we reject anew the doctrine that the Bishop of Rome should be able to pronounce, to establish and to prescribe infallibly as divine truth' any new doctrine 'not confirmed in Holy Writ nor universally confessed by the Church'.[158]

In January 1957 George Otto Simms was enthroned as Archbishop of Dublin at the age of 46; the Church of Ireland now had the oldest and the youngest archbishops in the Anglican Communion. In his sermon Simms declared that Anglican allegiance 'will not hinder but will rather help anything we can contribute in public service, in the field of education, in matters cultural and communal'.[159] Just how problematic that continued to be was thrown into relief by two events. A few weeks before the enthronement two young Irishmen had been killed during an IRA attack on an RUC police station in Co. Fermanagh, part of a renewed (though ultimately abortive) IRA onslaught on the 'Six

Counties'. Their funerals had been 'two of the most emotional in Irish history',[160] and were an illustration of just how difficult it was for Protestants in the Republic to feel a part of the national mood. The other event four months later was even more worrying. It happened at Fethard-on-Sea in Co. Wexford, where a mixed marriage was again the root of the problem. Mrs Cloney, an Anglican, decided to send her daughter to a Protestant school, and when the local clergy objected she 'fled' to Belfast with her children, refusing to return unless her husband would agree to their being brought up as Protestants. The local people rallied to the cause of the Roman Catholic husband, and – assuming that the local Protestants had had a hand in spiriting Mrs Cloney away – proceeded to organise a boycott. The local Catholic bishop declined to support his flock, though the Bishop of Galway spoke out against Protestants trying 'to make political capital, when a Catholic people make a peaceful and moderate protest'. Perhaps local memories went back to the original incident when in 1880 Captain Boycott gave his name to the practice. Perhaps the 1947 film, *Captain Boycott*, had inspired the people of Fethard-on-Sea.[161] Hubert Butler remembered driving the long journey to the village simply to purchase supplies from Protestant shops, in an effort to break the boycott, just as Orangemen had travelled to aid the original victim. Butler was scathing about the Anglican Bishop of Ossory, Phair, who had tried to play down the whole issue.[162] In fact it was de Valera, newly returned as Prime Minister, who spoke out eventually. 'I regard this boycott as ill-conceived, ill-considered and futile', he proclaimed, and 'I regard it as unjust and cruel to confound the innocent with the guilty ... I repudiate any suggestion that this boycott is typical of the attitude or conduct of our people.'[163] It was one of those 'moments of real dignity' to which de Valera could rise.[164]

The truth was that de Valera did not preside over a confessional state, any more that Brookeborough did in the north. The Protestant churches in de Valera's Republic cannot be compared with Protestant churches in Salazar's Portugal or Franco's Spain. Northern Roman Catholics were not treated like blacks in the southern states of the USA or in South Africa. In Eire the legal and constitutional framework might exclude Protestants emotionally, and sometimes – maybe during the Eucharistic Congress of 1932 or the papal visit of 1979 – a Protestant might feel like 'a stranger in a strange land',[165] just as a Catholic might feel in the north on the Twelfth of July. But the 1950s – despite Fethard-on-Sea or the IRA campaign in the border areas – did not see any serious destabilization of the Protestant churches. 'To enter the ministry in 1953', Bishop Warke recalls, 'was to become part of an edifice which had about it an air of changelessness.'[166] But things were about to change. When in 1959 Archbishop Simms presided at the General Synod for the first time as Primate, he was replacing a man whose first General Synod had been in 1915. 'Future writers', Archbishop Simms declared, 'may date Dr Gregg's resignation as the end of an era in Irish church History.'[167] It cannot have looked like the end of an era to Protestants who during the Patrician Year of 1961 were 'treated to "a glittering, bejewelled spectacle" as nine cardinals and over two hundred bishops and abbots arrived in Dublin for a triumphal expression of the faith'.[168] But underneath and beyond all of that, things were changing.

6
A Century at Home and Abroad

> Well may the leaders and members of the Church reflect on the awful seriousness of the simple fact that opportunities pass. It must use them or lose them. It cannot play with them or procrastinate to debate whether or not to improve them. Doors open and doors shut again. Time presses. 'The living, the living, he shall praise Thee'. It is the day of God's power. Shall His people be willing?
>
> *World Missionary Conference, 1910*[1]

Ecumenism

Tension between the Church of Ireland and the other Protestant churches did not disappear in January 1871; as we have seen, it took decades, perhaps longer, for the 'establishment' mentality to depart from the Church of Ireland, and for Presbyterians (in particular) to rid themselves of a feeling of second-class citizenship. Disestablishment had done much to rally the two churches to make common cause for the defence of Protestantism, when it had been the Roman Catholics who 'were siding, both theoretically and in practice, with the English Radicals and Dissenters'.[2] Those unlikely alliances, of Anglicans and Presbyterians on the one hand, and of Roman Catholics with English radicals and dissenters on the other, were cast into even starker relief by the home rule issue which, as we have seen, forced the Protestant churches to work together as never before. The Methodist newspaper noted in 1886 that their relations with each other 'were never so sympathetic as at present. Their unanimity of judgement and identity of attitude towards "Home

Rule" afford happy illustration of the solidarity of Protestantism, and herald a brighter future for their common work.'[3] But while the churches might coalesce in opposition to home rule, that 'brighter future' was still a long way off. The General Assembly declared in 1888 that 'the fortress of official favouritism is yet far from being dismantled, and further efforts will be required to raze it to the ground'.[4] When the Methodists in 1893 announced plans for a Dublin Central Mission, Anglicans attacked them, along with the Baptists and Plymouth Brethren, because 'these sects do not prey on one another; they find their quarry among ill-instructed members of the Church of Ireland'.[5] The Methodists were the target of Presbyterian irritation too; there was a fuss at the General Assembly in 1895 when the War Office was proposing 'that the Wesleyan Chaplain should for a time conduct Presbyterian services at the Curragh'.[6] Stronger Presbyterian fire power was directed at the Church of Ireland however. The issue was 'validity of orders', a sensitive topic for Anglicans since Pope Leo XIII's encyclical *Apostolicae Curae* of 1896, which had condemned Anglican Orders as invalid. Benson, Archbishop of Canterbury, happened to be in Dublin when the news broke, so Irish Anglicans were the first to hear his condemnation of this 'new instance today of defiance of history perfectly in accord with all we knew of Rome', and his admonition that this was 'a lesson to those who thought Rome open to argument'.[7] Presbyterians of course were not at all sympathetic to the Anglicans' determination to prove that their church maintained unbroken line of succession from the time of the apostles. For them, it was enough that clergy 'teach the Gospel Truth, and administer the Gospel Sacraments according to Apostolic rule and example'.[8] For them, the Anglican tendency to cast doubts on Presbyterian orders was obscurantist, and an obstacle to greater Protestant unity, because 'the theory of Prelacy with its priestly hierarchy is not evangelical, and bars all union'.[9] When in November 1900 the Anglican Revd R.W. Seaver spoke from the pulpit about Church of Ireland 'priests and people', he was taken severely to task by the *Witness*: 'We are not aware that the minister of St John's Church, Malone, and the other Episcopalian ministers are priests.' Then, rubbing a sensitive spot further, it went on: 'The Church of Rome disowns them and denies their priesthood. As there are no Protestant priests, the assumption is absurd.'[10]

All this, at the beginning of the new century, seemed a long way away from that 'brighter future' predicted in 1886; a long way too from the Presbyterian Moderator's confident prophesy of 1890. He declared that 'the next half century will see a great movement towards Union among the Churches ... I can never let go the hope that, were God's Spirit poured out on us abundantly, and all prejudice and party spirit sunk, the representatives of our great Churches which God has blessed about equally in their efforts on the mission field, would find that they agree more than they thought they did, and would by friendly consultation come to see things so far eye to eye as to discover some method by which our divisions might be removed.'[11] In fact, he was reading the signs of the times with some accuracy. Despite the squabbles between the churches, there were bigger forces pushing them closer together. Politics and the fear of Roman Catholicism were enormously powerful influences in the first decades of the twentieth century. But there were other factors as well, as we have seen, among them the growing common awareness of social problems, the temperance cause and the Christian Endeavour movement. All these were areas of common action, particularly among the laity. There was – even aside from the political pressure pushing them together – a widespread evangelical Protestant impetus in all the churches, evident in the support given to the Moody and Sankey Missions. When a Scotsman, James Macaulay, looked at the Irish churches in 1872, just after disestablishment, he anticipated that it would be the laity who would pioneer 'any movement towards the 'union and communion' which many desire'.[12] There was much truth in this. The clergy, by the very nature of their training, were conscious of the deep historical and theological divisions between the churches. The laity were more aware, by virtue of circumstance, of their common Protestantism. And by the end of the nineteenth century laymen figured largely in the church courts of the major denominations. So for a whole variety of reasons, the Irish entered what has been called the 'Ecumenical Century' more united than the English churches. There were also theological and ecclesiastical factors at work. Archbishop Crozier, who had just succeeded the aged Alexander as primate in 1911, spent much of his presidential address at General Synod that year expounding the 'keen and unquenchable desire for closer union and co-operation with our Protestant brethren all over Ireland'.[13] He quoted the resolution of the Lambeth Conference of 1908 which had called for more 'meetings for common acknowledgement of the sins of division, and for intercession for the growth of unity'. This, no doubt, had been the inspiration for his own initiative as Bishop of Down when he had arranged for meetings between Presbyterian and Anglican clergy to 'smooth the way for the coming union'.[14] He also made clear that the Protestant churches were feeling the positive effects of the Edinburgh Missionary Conference of 1910, which is often regarded as the beginning of the twentieth-century ecumenical movement. No doubt he also had in mind the Church of Ireland Conference in Belfast just six months earlier, when the Presbyterian Moderator had been applauded enthusiastically when he told the assembled Anglicans that 'the different branches of the one evangelical Catholic church of Christ are drawing closer together. I hope that movement will continue, and continue with accelerated speed.'[15] It wasn't all happy co-operation however, as the incident that blew up in 1913 illustrates. Provost Traill of Trinity, staunch churchman and landed Ulsterman, had offered the use of the chapel at TCD to the Presbyterians, then on second thoughts proposed that they use the hall instead for their service, but the Presbyterians insisted on going ahead with the original plan. Fellows who objected the Provost's action included J.A.F. Gregg, whose attitude to ecumenism was to be critical in the years ahead. Archbishop Crozier protested to Traill that he was guilty of 'giving over to the Presbyterians something which does not belong to you', but he also made the point that the affair was likely to 'throw back indefinitely the cause of Home Reunion, which the Bishops haven't much at heart'.[16] It was to be ten years before more formal steps were taken to further the 'Home Reunion' cause by the inauguration of Irish Council of Churches.

In those years the Protestant churches were thrust together by the political situation. They co-operated both at that level, and in other ways. In 1910 they were making common cause in Belfast against the incursion of Mormon missionaries. In 1911 they were joining in support of the Jews of Limerick, who were being subjected to harassment and victimisation as a result of fiery anti-Semitic sermons by the local Redemptorist priests. Their young people made common cause in the Irish Christian Fellowship, founded in 1913 as a graduate wing of the Student Christian Movement. Just occasionally the inter-church co-operation crossed the great sectarian divide. In 1911 the Anglican Bishop O'Hara of Cashel mounted the platform with the Roman Catholic Bishop Sheehan to launch an 'Immoral Literature Crusade' in Waterford.[17] More remarkable was a joint letter dispatched to the Prime Minister and others on 15 March 1915, in support of Lloyd George's crusade for stricter controls over the drink traffic. 'The country is more ready to submit to effective legislation in this matter than ever before',[18] the four main church leaders told Asquith, commending a scheme which was to be scuttled without trace by a combination of interests which included Irish members who feared for the brewing and distilling interests in their own constituencies. Still, it was a 'remarkable' letter.[19] But Catholic and Protestant leaders were soon at odds again, over the

Easter Rising and conscription. The Archbishop of Canterbury asked Archbishop Bernard in 1918 what he thought of the Irish Catholic bishops. 'I think very badly of them', he replied, adding that the RC Archbishop of Dublin 'is, I am convinced, proGerman'.[20] The events of wartime and the years immediately afterwards seemed to underline the 'struggle between two nations, the Protestant and the Roman Catholic'.[21] The Protestant churches now came together more formally than ever before. A resolution of the General Assembly in 1917 'noted with deep interest the appeal and suggestions in regard to the Church after the war, set forth by the Bishop of Down, Connor and Dromore in his letter to the "Spectator" of February 17th, 1917. The Joint Committee expresses its high appreciation of the spirit of the Bishop's declaration and its sympathy with the suggestions contained therein.'[22]

In 1919 the Bishop (D'Arcy) was welcomed at the General Assembly, and in his presence a resolution was passed asking their Committee on Co-operation with other evangelical churches to 'communicate with other Evangelical Communions in Ireland, with a view to co-operation'.[23] The Lambeth Conference of 1920 gave added impetus to these developments, with its *Appeal to All Christian People*. At the 1921 General Assembly, Gierson of Down, who had succeeded D'Arcy as bishop just in time to attend Lambeth, was a guest of honour. He addressed the brethren, and heard the motion which was phrased much more strongly than that of 1919, calling for a Committee which would confer not only on co-operation with other churches, but 'on the subject of Church Union'.[24] In November 1922 the Constitution of the 'United Council of Christian Churches and Religious Communions in Ireland' was adopted with members from the Church of Ireland, the Presbyterian Church, the Methodist Church, the United Free Church of Scotland, the Baptists, Congregationalists, Moravians, the Society of Friends and the Salvation Army. Co-operative action became more evident in the 1920s, not only in terms of the discussions at Council meetings, but also in the day-to-day life of the churches. The Presbyterian John Waddell preached at St John's, Malone, in 1923. Also in that year joint Methodist–Presbyterian services were arranged at Clonmel and twelve other southern churches, a co-operative venture which was to become a life-line for some of the scattered Protestant communities in rural areas. At the installation of the Revd A.W. Neill as minister of First Armagh Presbyterian Church in 1928, the Anglican primate attended; 'the first time probably in the ecclesiastical history of Ireland'.[25]

But beyond the realm of co-operation, there was the deeper question of organic union, which came to the fore in the Lausanne Conference on

Faith and Order in 1927, as well as in the ongoing discussions to create a united church in South India, and again at the Lambeth Conference in 1930. In his presidential address to the General Synod after the Lausanne Conference, Archbishop Gregg (presiding because of the illness of D'Arcy) rather gloomily confessed his inability 'to see how certain cleavages which became manifest there will ever be bridged'.[26] The Archbishop of Armagh, back in the presidential chair the following year, was characteristically more up-beat, but laid bare the crucial issue, that 'it is much easier to bring together the great Christian communions in matters of essential Faith than in matters of Order'.[27] The difference of approach turned on the question on episcopacy. Was the historic episcopate essential to the church, and in what sense? Was it of the *esse*, the very nature of the church, without which the church was not really the Church of Christ at all? Or was it of the *bene esse*, that is, the good order of the church, important to the church and historically part of it, but not – in the last analysis – crucial to its existence. Gregg himself, in this sense a traditional high churchman, was adamant: 'I view with great apprehension the tendency to advocate the acceptance of Episcopacy by those who tolerate it for the sake of reunion, on the explicit terms that no theory of the Episcopate, its origin, character or function, should be required of them.'[28] D'Arcy's stance was nearer that of Henson, the Bishop of Chichester, who shared with him his irritation after the 1930 Lambeth Conference. 'So long as episcopacy is looked upon as the *unum necesarium* of a Christian Church, I am sure that no reunion with the presbyterian and congregational churches is possible', he wrote, criticising the English primates for 'the intrinsic enormity of their assumption that the Christian fellowship is rightly linked and limited to episcopally ordered Churches'.[29]

Henson was right; it was on that very point that the subsequent discussions in Ireland would turn. Forty Presbyterian and Anglican clergy met in TCD in January 1931 to consider the question of 'Home Reunion'. The main speakers were J.S. Rutherford and T.W.E. Drury – respectively, leading Presbyterian and Anglican proponents of 'reunion'. The discussions bore fruit. In June 1931, the General Assembly set up a committee 'to meet a committee of the Church of Ireland to consider the subject of church union in Ireland, provided that, and so long as, the discussion is unrestricted and all relevant questions regarded as open, and, if possible, to suggest a scheme of union, or failing that, of co-operation'.[30] The General Synod resolved that a committee should meet with the Presbyterians 'to consider and to suggest a scheme for Reunion in Ireland'.[31] So the Joint Committee came into being, meeting first in

January 1932. In December 1933 an important resolution was proposed by the Presbyterian John Waddell, and seconded by W.S. Kerr, the Dean of Belfast, calling upon each church to declare that 'it fully and freely recognises, as a basis for further progress towards Union, the validity, efficacy and spiritual reality of both ordination and sacraments, as administered in the other Church'.[32] It was adopted by 24 votes to 3, Archbishop Gregg asking that his dissent be recorded. When Primate D'Arcy opened the General Synod in 1934, his address concentrated on the reunion question, and he threw his weight behind the interpretation of the Lambeth Appeal of 1920 as recognising the efficacy – the 'spiritual reality' – of non-episcopal ministries. He avoided, he said, the word 'validity' as ambiguous and legalistic. He ended his speech with a plea to go forward, 'not in any niggardly or grudging spirit', towards unity.[33] In fact the Synod asked for a suspension of the talks until the report of the conversations between the Church of England and Church of Scotland became available. The General Assembly, meeting a month later, adopted the December 1933 resolution, and recommended continuation of discussions on the basis that the Church of Ireland did the same, a 'precondition' that was going to be important. Gregg was not happy with the situation. At his own diocesan synod in October, he offered a different interpretation of the Lambeth Appeal to that of the Primate; he had attended the Lambeth conference in 1920, as he reminded the synodsmen with some asperity, and 'I am not without some reminiscences of what was intended by the appeal.'[34] This was Gregg's 'unyielding episcopalianism' which Henson of Durham had noted during the Conference. 'He is a much stronger Episcopalian than I am', Henson commented. No doubt Gregg would have been horrified by Henson's admission to D'Arcy that he had taken communion in a Presbyterian Church in Scotland.[35]

In January 1934, the 'Friends of Reunion' movement which had been founded in England was launched in Ireland, Drury and Rutherford prominent among its members. 'Reunion here', they felt, 'presents fewer difficulties' than in England, the churches in Ireland being already closer together.[36] Others felt the same. Colonel Madden told the Clogher Diocesan Synod that he was frustrated by the slowness of the movement towards reunion: 'for the life of him he could not see what was the difference between the Presbyterian church and their own Church of Ireland that would prevent them from uniting'.[37] Some were more cautious. The Presbyterian J.C. Johnston warned that 'several ministers have told me that these conferences were doing harm, raising false notions and unsettling our people'.[38]

The debate on reunion at the General Synod in 1935 was a stormy one, in the course of which the Archbishop of Dublin was effectively rapped over the knuckles by the Primate's assessor for an unprocedural intervention. Nevertheless, it was Gregg who carried the day by preventing the adoption of the resolution moved by W.S. Kerr, Dean of Belfast, that the Church of Ireland recognise 'the validity, efficacy and spiritual reality of both ordination and Sacraments as administered by the Presbyterian Church'. Gregg's counter-proposal, noting the 'condition prescribed by the General Assembly' in 1934, was that the whole reunion question must 'be deferred until more promising methods of approach present themselves'.[39] The Presbyterians in their General Assembly a month later resolved that the Anglican failure to recognise the 'validity of Presbyterian orders and sacraments' meant that 'further negotiations seem at present out of the question', and discharged the Committee on Church Union.[40] 'Much bitterness was created in Presbyterian circles', Professor John Barkley noted 35 years later.[41] It looked different from the Anglican point of view. Gregg, said one writer in 1963, had 'saved the Church of Ireland from what would have been a perilous lapse into latitudinarianism'.[42] That judgement was recorded just before a new round of talks began, in the very different world of the 1960s.

Recognition of each other's orders and sacraments was not a stumbling block between the Presbyterians and the Methodists. From 1937 a joint committee of the two churches worked together, on the basis of this mutual recognition, exploring 'whether closer union is practicable'.[43] They agreed on a definition of the ministry. 'Holy order comes, and must only come, by transmission, from those who have received transmitted authority to transmit', Gregg declared.[44] For non-Anglicans, that was a mechanistic and legalistic interpretation. They believed that 'the true Apostolic succession is that of Christian faith and experience ... Ordination is the public recognition of gifts of ministry and the giving of authority for their exercise. This authority is derived not from those already ordained, but from the whole body of the faithful in each denomination, and ultimately from God.'[45] The issue which terminated these Presbyterian–Methodist discussions was a much more practical one: the practice of itinerancy of ministers, 'to which the Methodist representatives attach very great importance', as the Presbyterians noted in 1945.[46] Given that no consensus could be achieved on this issue, the talks ended in 1947, though as one Methodist minister noted, 'the plain fact is that neither Church was really convinced of the need for union or ready for it'.[47]

This lack of enthusiasm for reunion and ecumenism was in a sense out of step with what was happening in the wider world. Just as the Methodist–Presbyterian talks were grinding to a halt, the Archbishop of Canterbury delivered a university sermon in Cambridge in November 1946, appealing to Free Churchmen to 'take episcopacy into their own system', though Fisher was more interested in 'intercommunion' than in 'organic unity'. In 1947 the Church of South India was inaugurated, with episcopal and non-episcopal churches organically uniting for the first time. In 1948 the World Council of Churches (WCC) was formally inaugurated in Amsterdam, and in the same year the Lambeth Conference recorded its thankfulness for 'the revival of interest in the cause of Christian unity which has been increasingly manifested in many parts of the world',[48] and although it did not suggest any new ways forward it greeted 'with satisfaction and hope' the 'proposals for organic union in various areas'.[49] The 1950s turned out to be what has been called a 'semi-ecumenical age', and this was true in Ireland as in England. But the decade that followed saw great changes, in this as in other ways. The Anglican–Methodist Conversations in England reached a critical stage in 1963. At the same time, there was an evident wind of change blowing through the Roman Catholic Church: the Second Vatican Council was opened by Pope John XXIII in 1962, and there was talk of *aggiornamento* (bringing up to date) in that seemingly most changeless of institutions. In Ireland, stimulus once again came from the Lambeth Conference, which in 1958 declared that to 'proclaim effectively the Gospel of God's reconciling love to the world, the Church must manifest in its own life the healing and reconciling power of the word it proclaims'.[50] The Lambeth committee on church unity had two members from Ireland, Bishops McCann of Meath and Mitchell of Down. The latter was regarded in Ireland as 'one of the most conservative, even "spiky", Anglican bishops'.[51] Yet it was he who, on returning from Lambeth, instituted a series of informal Anglican–Methodist–Presbyterian residential conferences at Murlough House, which 'prepared the ground for the later official Conversations between the Churches'.[52] A somewhat confusing array of inter-church discussions then began. The Presbyterians opened talks with the Congregationalists in 1961, and in 1964 the Methodists were included. Meanwhile bilateral discussions had opened between the Presbyterians and the Reformed Presbyterians. In May 1964, the Church of Ireland agreed to enter into conversations on a bipartite basis with the Presbyterians, while also opening separate conversations with the Methodists. The Congregationalists at this point withdrew from their talks with the Presbyterians. Three years later, these sets of conversations were amalgamated into a series of tripartite talks between Anglicans, Methodists and Presbyterians. An agreed 'Declaration of Intent' was drawn up at the first tripartite meeting in January 1968, which while 'seeking to preserve the truths in our several traditions', committed the three churches 'to seek together that unity which is both God's will and His gift', and invited other churches to join them.[53] The declaration was adopted by each of the churches at their annual meetings later that year. Symbolic of the new atmosphere was the attendance for the first time in 1971 of the Presbyterian Moderator and the Methodist President at the General Synod of the Church of Ireland.[54]

These tripartite conversations should be seen in the context of wider developments. There was increasing co-operation with the Roman Catholic Church. The Irish School of Ecumenics was founded in Dublin in 1970 by the remarkable Jesuit Michael Hurley, who was 'deeply conscious of the need for the Churches to break free from their sectarian history'.[55] There was increasing co-operation at grass-roots level, Presbyterians and Methodists sharing church buildings at Taughmonagh in Belfast in 1953 and then at Braniel in 1958. New shared buildings between Anglicans and Methodists were inaugurated at Glengormley in 1968 and Monkstown a year later. In 1972 the Methodists from Stephen's Green church in Dublin, which had been closed, began sharing with the Anglicans of Christ Church, Leeson Park. In Limerick a different sort of co-operation was instituted in 1973, suited to established congregations: the Presbyterian–Methodist Alternating Ministry scheme. Successive ministers come from the two different churches, and it was reported that the laity had taken to 'thinking of themselves as both Presbyterian and Methodist'. By 1999 there were six such 'alternating ministry' schemes in operation.[56] This mood of co-operation was dramatically symbolised by the Archbishop of Dublin when in October 1973 he stood in the dark and almost empty chapel of TCD, formerly set aside for exclusively Anglican worship, and declared it henceforth dedicated anew for 'Christian worship'.[57]

But no such gesture, and no such examples of co-operation, would bring about 'organic' unity among the Protestant churches. In the spring of 1973 there was anticipation that some such was imminent, in a 'scheme that would make possible a United Church with all the resources of a million-strong membership and more than 1500 clergy'.[58] The plan, called *Towards a United Church*, was commended for study at parish and congregational level, and the churches' governing bodies would assess the reaction and report back. A useful overview of what happened is given by the Revd Carlisle Patterson, who in 1958 became 'the first salaried ecumenical appointee in Ireland', when he was appointed part-time organising secretary of the United Council of Churches. He remained in post until becoming minister of Crescent Church in 1962, but continued to serve on the Council until 1970 when he took a position in England. He cannot conceal his disappointment that by then, the mood within Presbyterianism had changed and that (to use his bicycling metaphor) the church ' had passed a summit and was on an increasingly steep downward gradient'.[59] In 1970 the Presbyterians were concerned about membership of the World Council of Churches, and refused to contribute to its 'Programme to Combat Racism'. Pressures within the Assembly grew; it voted to suspend its membership in 1978, and by 430 votes to 327 the decision was taken in 1980 to withdraw. Perhaps, as has been suggested, this was due partly to the influence of the Revd Ian Paisley 'on the elders of the Presbyterian Church', or more generally because the 'debates took place in a volatile and highly charged atmosphere', given the IRA's campaign of violence.[60] It was in this climate that the discussions on *Towards a United Church* were taking place. Within the Presbyterian Church it was becoming clear that 'the majority of presbyteries were either lukewarm, apathetic, or hostile to the present Scheme, and to the general idea of Church union'. Patterson notes a feeling that during the 1970s 'the participants had grown disheartened by the apathy or hostility with which the products of their lengthy consultations had been received'.[61] Other commentators observed that the 'conversations have continued, but the heart has gone out of them'.[62] In 1988, the Presbyterians had gone out of them as well, leaving the Methodists and Anglicans to continue with a Joint Theological Working Party. In Ireland, as elsewhere, there was greater co-operation and understanding between the churches, but there was less talk of organic union in the final decades of the twentieth century. When the new Council of Churches of Britain and Ireland (replacing the British Council of Churches) was formed in 1990, it was only the Methodists and Anglicans in Ireland who joined. The Presbyterians by a two to one vote in their Assembly decided not to be involved with 'a Council which contained the Church of Rome whose doctrines were still contrary to those of evangelical Protestantism'.[63]

Missionary activity

In 1907 the Archbishop of Canterbury launched a book called *Church and Empire*.[64] It was a series of eleven essays 'on the responsibilities of

Empire', three of them contributed by Irishmen overseas. George Lefroy, Bishop of Lahore, wrote about 'Our Indian Empire'; M.R. Neligan, Bishop of Auckland, wrote on 'New Zealand – "An Ill-Constructed Quadrilateral"', and William Gaul, the diminutive Bishop of Mashonaland, contributed a chapter on 'South Africa: The Anglican Church and Imperial Ideals'. Ninety years later, in a volume entitled *'An Irish Empire'? Aspects of Ireland and the British Empire,* the editor noted that at a time when 'the substance of Empire has largely passed away, it may be that its most lasting legacy ... is chiefly intangible ... the traces which the British Empire has left in Ireland – and which Ireland has left in the British Empire – may most abundantly be found in social and cultural resonances, rather than in any concrete constitutional, political or even economic structures'.[65] We might want to add 'spiritual', for the Irish impact on the Empire and on the mission field in general was profound, and indeed continued long after the 'Imperial sunset'.

Irish missionary societies, or 'Hibernian' branches of British societies, flourished from the beginning of the nineteenth century. The Anglicans had the greatest proliferation. The Society for the Propagation of the Gospel in Foreign Parts (SPG), founded in England in 1701, had an Irish auxiliary from 1714.The Hibernian Church Missionary Society, founded in 1814, was an offshoot from the British body (CMS) which had been launched in 1799. The Dublin University Fukien Mission under the aegis of the CMS began work in 1886, and the SPG fostered the Dublin University Mission to Chota (or Chhota) Nagpur, started in 1891. There were Hibernian Auxiliaries of the South American Missionary Society (founded 1844) and the Colonial and Continental Society (1851). Women's (or 'Zenana') work was pioneered in India by the Gabbett sisters from Ireland working alongside the CMS in 1860, and Irish women were much involved in the Zenana Bible and Medical Mission (as it was named in 1880), and in the Society for Promoting Female Education in the East. The Countess of Dufferin, while her husband was viceroy, founded in 1885 the somewhat portentously named 'National Association for Supplying Female Medical Aid to the Women of India'. The Church of England Zenana Mission opened a Hibernian auxiliary in 1897. 'The Mission to Lepers in India' was formally launched in Dublin in 1878 by Wellesley Bailey, a teacher in the Punjab, where his experiences of leprosy moved him to return home to Dublin in 1873 and seek support for 'a Christ-like work' which he saw was needed. His friends the Pim sisters pledged support, and by 1889 the Mission had 26 stations in India and Ceylon. By the eve of World War I, the work had expanded into twelve countries, caring for some 14,000 lepers. By 1962 it had became a world-wide and inter-denominational organisation, renamed the Leprosy Mission.[66] It retained its spiritual basis, advertising its work in 1999 as '125 Years of Care and Prayer'.

Irish Anglican efforts were particularly focused on India, as is testified by the number of Irishmen who became bishops there in the first half of the century: R.S. Copleston (Colombo, 1875–1903, Calcutta 1902–13), G.A. Lefroy (Lahore 1899–1913, Archbishop of Calcutta 1913–19), E.A. Copleston (brother of R.S., Colombo 1903–24), Eyre Chatterton (1st Bishop of Nagpur 1903–19), H. Pakenham-Walsh (Assam 1915–23), K.W.S. Kennedy (Chota Nagpur 1926–36), A.O. Hardy (Chota Nagpur 1937–48) and F.R. Willis (Delhi 1951–66). It was a significant succession. The Chota Nagpur mission was founded by young TCD graduates, following the example of the Oxford mission in Delhi and the Cambridge one in Calcutta. Echoing the Student Volunteer Missionary Union's motto of 'The Evangelisation of the World in this generation', they proclaimed in their appeal for volunteers 'that Christ wishes the whole world to be evangelised, and in our time has wonderfully opened the way for this to be done'.[67] The first volunteers formed themselves into a 'brotherhood' under the auspices of the SPG, and sailed from Tilbury in December 1891. Among the young clergy aboard were the future bishops Eyre Chatterton (first head of the mission) and K.W.S. Kennedy, and a nurse, Fanny Hassard. Their destination was a former military station in Chota Nagpur, a diocese in north east India which was roughly the size of Ireland. They employed the classic threefold missionary thrust of preaching, teaching and healing. They soon realised that 'the preaching of the Gospel in India must be done by Indians'[68] and the first local priest, L.P. Singh, was ordained in 1908. By 1918 there were 44 schools, as well as St Columba's College, which became the first tertiary educational institution in Chota Nagpur. The medical work was 'a principal source of evangelisation', and if there was any resentment at the Irishmen's conversion activities, 'their medical services practically cooled it down'.[69] The cost soon became clear: 1912 saw the deaths both of the head of the medical mission, Dr Heard, and of the pioneer Miss Hassard. The lesson was spelt out back home in Ireland: 'It will be an evil day for the Irish church if her young men and women hang back in sloth, and refuse to help the spread of Christ's Kingdom throughout the world, by personal service abroad, by constant prayer and work at home.'[70]

Far down in the south of India, a young Irishwoman was engaged in a missionary enterprise which, although not sponsored by the Church of Ireland, was an inspiration to many of its members, and to non-

Anglicans as well. Amy Carmichael was a Presbyterian who as a young women was instrumental in setting up 'The Welcome', a hall in Belfast which was the base for her work among the 'shawlies' or mill girls of the city. Although dedicated in 1889 by her own minister, it was to be an inter-denominational evangelistic centre. While Amy left soon after it opened, the hall has its own place in Irish missionary activity, with what its historian calls 'a vision for the evangelisation of the uttermost parts of the world'.[71] It sent out a missionary in the 1920s to work with C.T. Studd in the Congo, and another who laboured there for 30 years up to 1964. The foundress, meanwhile, had tried her hand at missionary work in Japan, and then in 1895 was accepted by the Church of England Zenana Missionary Society, starting work the following year at Dohnavur, in the Tinivelly district of south India. Her endeavours led to the creation of the Dohnavur Fellowship, which became a non-denominational enterprise after 1925, when the links with the CEZMS were cut. She herself saw no need for denominational narrowness in her work. She was happy to quote from Bunyan, or Julian of Norwich or Thomas à Kempis;[72] she was amused when her inviting Bishop Tubbs of Tinivelly to dedicate her 'House of Prayer' in 1927 caused 'hesitation in the minds of some of her friends [in Ireland and England] concerning Dohnavur's intercourse with Bishops!'[73] Those same friends would have approved of the fact that 'she constantly read stories from Foxe's *Book of Martyrs* to the older girls',[74] though one wonders what the Indian girls made of it. She remained at Dohnavur until her death in 1951, and long outlived her sister Eve, who for 40 years from 1896 was working for the South Africa General Mission. That one family should produce two such long-serving missionaries was not uncommon, such was the enthusiasm for the 'evangelisation of the world'. A glimpse of the candidates for the mission field can be gained through the minute books of the Hibernian Church Missionary Society. Miss Maria Bluett volunteered in 1899, and made a favourable impression, for despite her 'somewhat advanced years' she would 'practically be an Honorary Missionary, having promised to contribute £100 a year towards her own support'.[75] The case of Samuel Marsh in the same year was deferred, for although he was recommended by the rector of Coalisland, he was only 18.[76] In 1900 Mr W. Sweetnam from Skibbereen was recommended to London as a missionary candidate, even though handicapped by the fact that his family 'attended services in the Methodist Church'.[77]

Irish Anglicanism's most famous missionary son was by that stage already dead, and the memory must have been fresh in the minds of those candidates. George Pilkington was born in Dublin in 1865, the son of a QC who was involved in the framing of the constitution of the church when it was disestablished. George was educated at Uppingham School, and in 1884 went up to Pembroke College, Cambridge. The Moody and Sankey mission of a few years earlier had made quite an impact, and had inspired C.T. Studd and six others (known as the 'Cambridge Seven') who were to become missionaries with the China Inland Mission in 1887. They 'greatly quickened the Missionary spirit in the "Varsity"',[78] and Pilkington arrived in time to hear them speak. It was in this atmosphere that he had a conversion experience, joined the CMS, and sailed for Africa in 1890. It was to be a short but enthusiastic ministry in Uganda. He plunged into learning the local language and translating the 1662 Prayer Book; he prepared a Bible translation; he taught other volunteers, acted as a translator and was a successful preacher both in Uganda and back home on furlough. He was killed in December 1897, joining a distinguished list of Anglican martyrs there, from the 32 boys murdered by the pederastic king Mwanga in 1886, to Archbishop Luwum who was assassinated, most likely on Idi Amin's orders, in 1979. Irish Anglicans could look elsewhere, too, for inspiration from martyrs. The Dublin University Fukien (or Fuh-Kien) Mission was inspired by the Dubliner the Revd R.W. Stewart, a CMS missionary. From 1888 it raised funds and sent out graduates to the province of Fuh-Kien, a coastal region north of Hong Kong, which was the most thriving area of Anglican missionary witness. The founder and his wife were to be massacred there in 1895.

In 1919 there were 141 Irish missionaries serving with the various societies of the Church of Ireland; of these, by far the largest numbers worked in India (72) and China (64).[79] Fifty years later the total had fallen to 73, of whom 14 worked in India or Pakistan, and one worked in Hong Kong with what was now called the Dublin University Far Eastern Mission.[80] A century which had begun with a widespread desire to evangelise the world, and with men and women prepared to pay the ultimate price for that work, was ending rather differently. It was a phenomenon evident in the missionary work of the other churches as well. The Presbyterians began the century with 63 missionaries, of whom 33 were in India and 20 in China. Unlike their Anglican contemporaries who went out under the aegis of various missionary societies, Presbyterian missionaries were commissioned by their church. The very first Irish General Assembly in 1840 had ordained two men for work in the Gujarat peninsula in north west India, and that was to become the major mission field for Irish Presbyterianism. Here again there was a threefold thrust of preaching, teaching and healing. Medical work had been confined to dispensaries until the beginning of the twentieth century, which opened with famine and a cholera epidemic. The charge that the church took the opportunity presented by these disasters to increase the numbers in the Christian community has been ably countered,[81] but certainly in its wake the Presbyterians increased their medical input, the first hospitals opening in 1905 and 1906. Three others followed in 1919, 1936 and 1939. Thanking the Boys' Brigade for its efforts in raising money for the operating theatre of the Alexander Kerr Memorial Hospital (opened in 1936) and calling for more funds from Irish Presbyterianism, the medical superintendent wrote: '... you may well ask, what will the return be? Our reply is, the blind will be made to see, the lame to walk, and the sick will be ministered unto, and the poor, yes, and the rich, will have the Gospel preached to them. I know no better return for money that can be had in this present day.'[82] To 'have the Gospel preached' had always been the primary aim. The local language was important to that end, and the Bible had been translated into Gujarat in 1829. Local elders were first ordained in 1875 and the first Indian minister in 1888. The new century opened with the inauguration of an autonomous 'Presbytery of Gujarat and Kathiawar', which four years later became part of the newly created Presbyterian Church in India, although the Irish missionaries still controlled much of the overall strategy and property. For the work of evangelism, theological education was a necessity, and Stevenson College in Ahmedabad was opened in 1892. There was a network of schools, both Gujarati and English medium, and there were training centres, especially those created during the famine to provide a means of subsistence for the local poor.

Like the Anglicans, the Presbyterians found their enterprise costly in terms of human life: 21 Irish ministers are buried in India. The cost to native Christians could be vastly greater, as it was in China, the other focus of Presbyterian mission. Over three hundred Manchurian converts were murdered during the Boxer Rising which opened the century. The missionaries escaped with their lives. There had been an Irish Presbyterian presence in that part of the Chinese Empire since 1869, but it expanded hugely in the 1890s. By 1900 there were nine Irish mission stations and 93 outstations, with their centre at the capital, Moukden. Pre-eminent among those who served there was Thomas Fulton. A member of the choir of the old Fisherwick Place Church in Belfast in the 1870s, he had responded to the collection at a foreign mission service by placing with the gifts a slip of paper on which he had written one word – 'Myself'. In 1885 he was ordained to the mission field, and laboured in Manchuria for over half a century, until World War II and old age caused him to retire in 1941.[83] Despite the huge political upheavals of those years, the indigenous church in Manchuria developed towards its goals of self-support, self-propagation and self-government.[84] It did so earlier and more speedily than the Church in India. Already in 1891 the missionaries had set up a Presbytery in Manchuria, involving Chinese elders and using Chinese language. Alongside this the missionaries organised themselves into a Conference, which included all the Scottish and Irish missionaries and their wives, and after 1916 the Danish missionaries working in the area. When the Presbyterian Church of China was formed in 1907, the Manchuria Presbytery, with the consent of the Irish and Scots churches, became part of it. When the missionaries' own conference discussed 'independence' in 1911, it concluded that the Manchurian Church was already independent by virtue of its membership of the Church in China. No formal decisions were made in Belfast or Edinburgh until 1922, when a deputation from Ireland and Scotland visited Manchuria. It accepted that the missionaries were, in effect, the servants of the local church and not of their home churches, and agreed that policy should be formed within the Synod – which by then had twice as many Chinese members as European. It had been a rapid movement towards what would later become known as indiginisation.[85]

Just as the Anglicans fielded far fewer missionaries in 1969 (73) than in 1919 (141), so too Presbyterian numbers dropped in the same period, from 64 to 52. The drop was much less dramatic, partly because a new field had opened up. In the late 1950s, the Irish Presbyterians joined their Scottish brethren in Nyasaland (Malawi) where Presbyterianism was strong, and by 1969 there were 15 missionaries from Ireland at work there. But by 1975 the total number of Presbyterian missionaries had dropped to 33, which was, like the Anglican numbers, about half the 1919 total, and by 1990 the total was halved again. Clearly the great days of missionary endeavour were over. What had happened? In the case of China, the second most important area of operations for both Presbyterians and Anglicans at the beginning of the century, the Communist takeover is enough the explain the foreclosure of the missionary effort. In August 1949 when the Presbyterian Conference of missionaries met in Moukden; 'some, at least, were doubtful of the value of the work they were being given to do and some were convinced that their very presence was now proving to be an embarrassment to their Chinese friends and colleagues'.[86] It was the last time the Conference met and in August 1950 the remaining missionaries left the country. The

Dublin University Fukien Mission faced similar problems. By 1950 most of the missionaries had left or were in the process of withdrawing to Hong Kong, and the re-named the Dublin University Far Eastern Mission concentrated its personnel in Malaya and Singapore.[87] Such dramatic political upheaval did not hit the work in India. From the mid-1950s onwards there were strong local moves to outlaw proselytism, though the bill which was introduced in Gujarat in 1972 was declared *ultra vires*.[88] In 1968 the Presbyterian Foreign Missions Board noted the Indian government's policy that 'the case of each missionary will be considered individually and "Indianization" of all Christian missions will be reached on a "progressive" basis', though the Board was confident that the Indian government would not move to expel all foreign missionaries.[89] Presbyterians and Anglicans in north India were then faced in 1970 with a unique situation – the creation of a united church into which they both were entering, with the blessing of their respective parent churches. They had entered the Church of South India in 1947, but that had not involved Irish missionary personnel. Now in 1970 the Church of North India was inaugurated, in the area in which the Irish missionaries were concentrated, and with the full blessing of the Irish churches. In its last message in January 1970, the (Anglican) Church of India Pakistan, Burma and Ceylon asked the other uniting churches 'for forgiveness for our share in the sin of division in the past', and assured the other churches of the Anglican Communion that it would 'take into the United Church its full spiritual heritage of faith and order'.[90] It was a message that must have struck chords with Anglicans and Presbyterians at home, launching their own unity discussions at that time. Perhaps it was a portent for those discussions that the General Assembly's debate on the Church of North India lasted three and half hours, and was 'somewhat acrimonious', the nub of the problem being the episcopal nature of the new church.[91] By the end of the twentieth century it still looked unlikely that the Irish churches would see such an appointment as now took place in India: Donald Kennedy, ordained to the mission field by the Irish General Assembly in 1942, became the Bishop of Bombay in the United Church in 1974. Thus Irish involvement in India continued, though in 1990 Dr Jean Shannon 'the last serving Irish missionary in Gujarat, returned to Ireland on final furlough'[92]– at the end of a long succession of some 300 Irish Presbyterians who had served there in the previous 150 years. Robin Boyd, himself a missionary there from 1951 to 1975, has identified four reasons for 'decline'. First, there was the policy of transferring leadership into Indian hands; second, the more restrictive policies of the Indian government; third, financial crisis with rising pay scales and large overheads; fourth, the family pull, whereby missionary parents felt they needed to return to home to educate their children. But there were deeper forces at work, as a church commentator in Belfast noted in 1973. He identified 'collisions with nationalism' and also a failure of nerve within the various missionary headquarters, as a result of 'the disappearance of the old certainties which grew and flourished in the days of the now-vanished British Empire'.[93] The WCC's World Conference on Church and Society in 1967 illustrated the new mood. Some 350 participants (including just one Irishman, David Bleakley) gathered in Geneva to discuss 'Christians in the technical and social revolutions of our time'. There was no mention of missionary work.[94] The 1973 Bangkok Assembly of the Commission on World Mission and Evangelism was specifically focused on 'missionary work', but its language was far distant from 'the evangelisation of the world in this generation', when it spoke of 'the struggles for economic justice, political freedom, and cultural renewal as elements in the total liberation of the world through the mission of God'.[95] The WCC Nairobi Assembly two years later spoke the same language, though it did call for Christians to 'recover the sense of urgency' and to 'proclaim the saving word of God – today'.[96] The question was: how? The Reith lecturer in 1979, looking back on that Assembly, was scathing, claiming that Christianity was being reinterpreted as 'a scheme of social and political action, dependent, it is true, upon supernatural authority for its ultimate claims to attention, but rendered in categories that are derived from the political theories and practices of contemporary society'.[97] Certainly for the Churches in Britain and Ireland, the focus in the last quarter of the twentieth century had shifted towards a deeper involvement in the social, political and medical needs of men and women, especially in the Third World. A major section of the Church of Ireland Standing Committee's report in 1999 dealing with overseas work concentrated on the £279,325 of grants in aid to development projects world-wide.[98] The papacy, meanwhile, employed language which was closer to traditional Protestant concerns. The 1994 Encyclical *Veritatis Splendor* admonished the faithful: 'Do not be conformed to this world', and called them to 'Redemption' and 'new life', whereby people could be 'freed from the slavery of evil and given the ability to do what is good'.[99] The Presbyterians finished the century employing language of a similar nature. The General Assembly of 1999 was reminded 'that there are millions in our world today who are lost for all eternity because they have not heard or have not listened to the Gospel of salvation through Jesus Christ. We are called to go forth and make disciples of all nations to the Glory of God and the extension of his kingdom.'[100] It was a call which would not have sounded out of place in the General Assembly of 1899.

The missionary activity of Irish Protestants in the twentieth century was not confined to India and China. The clergy of the Diocese of Connor, for instance, who at some stage served on the mission field in the course of the twentieth century, numbered some 47 (accounting for just 4% of all clergy who served in the diocese). Of those, 17 laboured in Africa, 15 in India, 10 in the Far East, 4 in South America and one in the Middle East.[101] The Presbyterian Church found a new outlet for missionary activity in 1979 via a link-up with the United Mission to Nepal – a country only opened to foreign mission in 1954. By 1999, ten missionaries (or couples) were at work there. The Methodists in the late 1960s had 35 missionaries in the field, an astonishingly large number for so small a church, given the Anglican and Presbyterian figures of 73 and 52. The Irish Methodists' greatest period of influence had been in the late eighteenth and early nineteenth centuries, when they supplied pioneer workers opening up the Wesleyan missionary endeavour in Newfoundland, Ceylon, Madras, Cape Colony, Mexico, St Kitts, Jamaica, Bermuda, Gibraltar and Hong Kong, though by far their greatest feat was the planting and promoting of Methodism in the newly independent United States of America.[102] Its missionary enthusiasm for the rest of the nineteenth century seems to have been at a low ebb, though rekindled thereafter with an awareness that it 'would increase, not lessen, the effectiveness of witness within Ireland itself'.[103] A Presbyterian observer noted that there might well be a growth in two-way traffic of another kind as well, in terms of 'a large increase in the number of Third World Missionaries working in the British Isles', even if this might not yet affect Ireland, 'where the major churches are still strong'.[104]

An examination of the missionary work of the churches forces one to look outwards, away from Ireland's shores. Western churchmen in general needed a similar corrective. The American Episcopal Church was being warned in 1992 that the 'preoccupation with our own debates over sexuality, authority and language, have blinded us from the joys and struggles of our sisters and brothers in Christ around the world'.[105] The world, as the millennium closed, saw huge Christian growth in Africa, and Islamic expansion with a world-wide growth rate twice that of the Christian churches. At the 1998 Lambeth Conference, both of these wider perspectives were much in evidence. 'The power in the Anglican Communion has, indeed, shifted southwards', it was noted, as white Anglo-Saxon bishops no longer outnumbered others. It was also noted that for the first time, Islam was on the agenda.[106] For Anglicanism, and for the Irish churches, that was a useful corrective to parochial concerns and local issues. Another wider perspective was available to Irish Protestants, because for some two centuries they had been travelling abroad, not only as missionaries, but as emigrants.

Emigration

When the nineteenth century closed, there were some 4.5 million people in Ireland – and some 2.5 million Irish-born people living overseas. If, as has been claimed, 'Protestants migrated from Ireland in close to the same proportions as did Roman Catholics',[107] then there were about 600,000 Irish-born Protestants living outside Ireland. Those who could claim Irish Protestant descent far outstripped that number. Around 1.5 million had emigrated since the first significant wave in the late eighteenth century, when Presbyterians had fled persecution in Ireland and had played a key role in the emerging United States of America. All these emigrants were part of a larger phenomenon, part of 'a great diaspora of the English-speaking people'.[108] There was, as George Kitson Clark put it, 'a whisper in the ear of people in all ranks of society ... "unless you wish never to improve your lot, unless you are content going to sink into a lower level, you must go, you must go, you must go". Many listened to it, the intelligent labourer, the younger sons of small farmers, the graduates of Glasgow, Edinburgh and Trinity College, Dublin, who saw little chance of fruitfully employing their talents as medical men, lawyers or divines in the crowded land of their birth.'[109]

The Protestant population of Dublin City fell from 45,896 in 1911 to 27,506 in 1926, a 40% decline. 'Though emigration has been treated as peculiarly characteristic of Catholic Ireland', writes a commentator on those figures, 'it is probably true to say that Protestant Ireland is a population more shaped and conditioned by emigration'.[110] The Methodist population of Ireland as a whole experienced a 38% decline from 1844 to 1900, and some 33,000 emigrated. In the midst of that seepage the Methodist Conference lamented: 'The tide never ceases ... it is always an ebbing tide ... and the loss is all the greater as our emigrants consist chiefly of the young and energetic.'[111] So 'need we wonder', said the church newspaper, that Protestanism 'should be but in struggling condition, when its very best life blood is pouring into other countries'.[112] They poured all over the world; to Britain itself, but also to the United States and Canada, to Australia, New Zealand and South Africa. By the turn of the century nearly all the Presbyterian emigrants from the once prosperous Braid Valley in Mid-Antrim were going to Canada. It was said that 'some Toronto department-store and factory owners virtually guaranteed employment to loyal Protestants born in Ballymena'.[113] In 1895 the General Assembly asked its 'Committee with reference to Emigration' 'to make what arrangements may seem best for giving information, counsel, and friendly, but not financial, help to emigrants to Canada'.[114] So although emigration was draining the 'life blood' of the churches, they saw it as their business to give help and counsel to such emigrants, and that meant maintaining contact with churches overseas, as well as providing chaplains at the ports of departure. The Dublin-based Association for the Relief of Distressed Protestants saw clearly the dilemma, for while 'emigration from this country is not good', it was sometimes 'the only remedy and sometimes it is a very real and permanent form of relief to send a man to Canada'[115] – north west Canada was the favoured destination for emigrants aided by the Society in 1907. When Mrs May Edmundson, a long-time supporter of the Dublin Prison Gate Mission, died in 1906, there was a proposal to 'raise a permanent memorial to her'. Given her interest in 'the emigration of young girls', it was decided that the most fitting memorial should be a fund which would help deserving cases.[116] World War I put a temporary halt to emigration, but the 1920s saw a further rise in the numbers departing 'from our shores to the colonies, and especially to Canada'. The Presbyterians again mobilised their resources, though with a heavy heart: 'There is hardly a congregation throughout our bounds which is not losing some of the very best of its members.'[117] The Colonial Mission committee arranged for chaplaincy work at ports of departure, aboard ship, and at ports of disembarkation. A handbook called *Immigrant Conditions in Vancouver BC* was published to help emigrants; it included advice from Canada that 'a church connection is of the greatest importance to a family man here'.[118] The Anglicans too were anxious that their emigrants should know that their 'Port Emigration Chaplain' in Belfast was on hand with advice and counsel.[119] Emigration was less common during the years of the Great Depression, and indeed there were reports of people returning from overseas. The Presbyterians were again a bit ambivalent. They might lament the loss of church members from the home congregations, but they were also aware that they had 'a vital contribution to make to the nations of their adoption and a witness to bear to the Reformed faith'.[120] The General Assembly was told in 1934 that while emigration had 'been practically at a standstill', the fact was that in Australia and other overseas territories there were vast spaces crying our for settlement: 'One trusts that soon a movement may commence to plant these vacant spaces with settlers of our own kith and kin.'[121]

Not only did ordinary church members leave the Irish churches to settle overseas; so too did the clergy. A calculation from some available sources shows the nature of some Baptist and Anglican clergy emigration in the course of the twentieth century (Table 6.1).[122]

Table 6.1: Baptist and Anglican clergy emigration in the twentieth century

	Baptist Ministers	*Connor Clergy*	*Down & Dromore Clergy*	*Totals*
Total	280	1172	810	2262
Total emigrating	42 (15%)	251 (21%)	127 (16%)	420 (19%)
Emigrated to:				
England	20	212	99	331
Scotland/Wales	9	4	3	16
Canada	9	14	9	32
Australasia	2	15	5	22
USA	2	6	3	11
Elsewhere	–	–	8	8

This very imperfect snapshot shows that nearly one-fifth of all these Irish-born clergy ended up resident outside Ireland, mostly in England. There was an outflow too from Irish Presbyterianism, which in the 130 years after 1842 provided about 1,200 ministers to work in their sister church in England.[123] Very few of the Connor clergy left the priesthood – about 5 in the course of the century, while 47 served as missionaries overseas at some stage in their ministry. The Baptist clergy were less stable as a body: around 16 of them (5%) left the Baptist ministry to enter the Presbyterian church (6) or other churches as clerics. But of the Irish clergy who emigrated, the Irish Anglicans are the easiest to trace, especially those who became bishops overseas. In the course of the century, Irishmen took over the dioceses of Niagra, Montreal and Brandon in Canada, Aukland in New Zealand and George in South Africa. In England Bloomer of Carlisle (1946), Greer of Manchester (1947), and Williamson of Southward (1991) were full diocesan bishops. Assistant and suffragan bishops included Walsh (Ely 1942), Hardy (Bradford 1948), Coote (Fulham 1957, Colchester 1966), V.J. Pike (Sherborne 1960), St J. Pike (Guildford 1963) and Richmond (Repton 1986). One of Ireland's leading churchmen, T.C. Hammond, Superintendent of Irish Church Missions from 1919 to 1936, emigrated to Australia where he was Principal of Moore College (1939–61) and a leading evangelical in the very evangelical diocese of Sydney. Presbyterians too moved overseas, the link with the USA symbolised by James McCosh, who was President of Princeton from 1868 until his death in 1892. The Australian link is personified in J.D. McCaughey. He ordained in Ireland in 1942, went out to an academic post, became President of the Uniting Church in Australia in 1977–9 and was appointed Governor of the State of Victoria in 1986.

The overall scale of Irish emigration, while lower in the twentieth century than in the nineteenth, was a continuing cause for concern among the churches, and especially in terms of the diminishing Protestant population in the south. A Jesuit priest examining the statistical evidence in 1945 declared that 'the Episcopalians are in present jeopardy'.[124] He was commenting on an Anglican pamphlet which gloomily recorded the statistics.[125] Such evidence is, however, a notoriously unreliable predictor. As Paul Compton has pointed out, if the statistical trends of the 1880s had been extrapolated, the Roman Catholics in Northern Ireland would by 1981 have been a mere 15% of the population.[126] Nevertheless the worries were there. The Presbyterians and Anglicans, as we have seen, were deeply disturbed by the decline of their numbers in the south after partition. Archbishop Gregg's first presidential address to General Synod in 1939 focused on the problem. Why a 12% decline in just ten years up to 1936? He concluded that it was not so much emigration, nor even mixed marriages, that were to blame, but rather late marriage, or no marriage, among his southern flock.[127] The mixed marriage issue was to recur again and again. Protestants did not forget the *Ne temere* scare of 1910. And yet Gregg may have been right. A Methodist questionnaire in1938 revealed that 'the menace of mixed marriages had cost the Methodist Church practically no members over a period of 25 years'.[128] Bishop Harvey at the Cashel Diocesan Synod in 1939 also identified emigration and the postponement of marriage rather than mixed marriages. 'As grass grows', he said, 'and water flows so will young men and women go where a prospect of a fuller life and better remunerated work awaits them.'[129] The argument about the causes of population decline would continue. An English Roman Catholic sociologist told the 1972 Social Study Conference at Falcarragh that 'the mixed marriage laws are gradually throttling the Protestant community in Ireland',[130] while Garrett Fitzgerald told a consultation at the Irish School of Ecumenics in 1983 that 'as a result of mixed marriage, an erosion of the Protestant population by about 25 per cent had occurred'.[131] Yet in that same year a reporter produced the startling statistic that the 'children of 80% of mixed marriages in one diocese of the Republic are being reared in the Protestant churches', and that 'in some other dioceses over 50% of such children are Protestant'.[132] In 1996 Bishop Darling declared that many couples in a mixed marriage were 'tending to go the Church of Ireland way'.[133] Whatever the reasons for decline, Gregg's three factors seemed relevant: mixed marriage, emigration, and late marrying, probably in ascending order of importance.

Emigration continued. The Methodists produce annually the most detailed and comprehensive statistics, and it is instructive to compare the second half of the nineteenth century with the second half of the twentieth (Table 6.2).

Table 6.2: Emigration of Methodist Church members, nineteenth and twentieth centuries

	Nineteenth century		*Twentieth century*
1850–9	6,811	1950–9	2,333
1860–9	5,340	1960–9	1,802
1870–9	3,682	1970–9	1,534
1880–9	4,717	1980–9	780
1890–9	5,725	1990–9	500 (estimate)

So, despite 'the Troubles', the problem is far less than it was. The headline in a local paper in 1971 was: 'Shock Ulster Exodus is revealed', and it is true there was a small rise in 1971, but the overall figures did not warrant any great 'shock'.[134] Nevertheless there were two issues related to emigration which caused concern for Protestants in the late twentieth century.

One cause of concern was the exodus of young people, particularly to England. It was an old problem. In the 1870s, for instance, far more Irish boys crossed the sea to attend boarding school in England than were boarders in Irish schools.[135] By the end of the century Kilkenny College – *alma mater* of Archbishop Magee and of Earl Beattie – was reduced to one pupil. (A century later it had 500 boarders and 250 day pupils.) Maurice Hime, the headmaster of Foyle College, Londonderry, wrote extensively on the problem, pointing out that the exodus was a financial loss to Irish education, fees being invested in English schools rather than stimulating growth and development of Irish ones.[136] It was a loss, too, of talent. Boys were sent to England to make the right connections, to enhance their job prospects, and 'to cure their brogues'. Basil Brooke and his brothers were sent to England for that reason, and consequently, we are told, they had 'singularly pleasant speaking voices'![137] James Craig, however, was sent to school in Scotland, where 'he heard in Edinburgh accents that were not far removed from his own', with the result that he never puzzled his Ulster constituents 'with an acquired accent or unfamiliar phrases'.[138] This is a matter of more than aesthetic importance, as the Revd W.F. Marshall was aware: he became a protagonist for the Ulster accent in the 1930s, criticising the BBC for its rejection of regional accents and the consequent 'disposition to deny an Ulster identity'.[139] Prime Minister Terence O'Neill, and his Home Affairs minister, Captain Long, had their 'patriotism' called in question in 1969 on the grounds of 'possessing a non-Ulster accent'.[140] Accent-related identity was a more central issue for Presbyterians, and the smaller Ulster denominations than it was for the Church of Ireland, which still clung to its identity as a national church, and retained its headquarters and its clergy training in Dublin. But identity was an issue. So too was the question of promotion prospects. Provost Mahaffy of Trinity was lamenting in 1895: 'No success in an Irish school, no training as a scholar, no proficiency as an athlete, will obtain for an Irishman coming from Ireland any fair consideration.'[141] Little wonder that the graduates of his university listened to that voice which Kitson Clark spoke of – the voice saying 'you must go'. The situation at the end of the twentieth century was different. There was no longer the colonial Empire in which to settle; there was no longer a prejudice against local accents within the United Kingdom; and – most importantly – there were 'the Troubles'. A survey of Ulster schoolboys in 1971/2 suggested that two-fifths of the secondary pupils and a third of the primary ones 'wanted to leave Northern Ireland when they are older'.[142] Most would, if they fulfilled their ambition, go to Britain, and such movement does not show up in emigration statistics, but it was still the 'brain drain' that had afflicted Ireland for nearly two centuries. Garrett Fitzgerald warned in 1998 that while less than a quarter of pupils from Northern Ireland's Roman Catholic schools went on to higher education in England, 46% of Protestant pupils did so, and of those, only 29% returned to the Province after graduation.[143] In the early 1990s the establishment of 'Youth Link NI' – a unique inter-church youth organisation with substantial government funding – had as its core aim the development of links between the youth of the four main churches. 'For the first time', it was reported, 'the churches are organised together for peace, better understanding and community relations and development.'[144] It remained to be seen whether such an initiative could give young people a desire to stay in Northern Ireland and work together.

There was another related concern. Among the Protestants of the North was the growing belief that they were being 'outbred'. It was not a new concern. Lord Charlemont, Ulster's second Minister of Education, wrote – somewhat sardonically – in 1933: 'In another couple of generations the Protestant majority in Ulster will have vanished so far as I can see – and then there will be really jolly times which I am glad to think I will not see.'[145] Sixty years later a Catholic commentator was drawing attention to the current 'crude consideration of how narrow the majority's margin is, how soon it will be necessary to stop describing Catholics as a minority'.[146] The 'shifting demographic tide', as one outside observer called it,[147] seemed to be favouring the Catholics. Marianne Elliott noted in 1992 that 'current demographic trends are contributing to an apocalyptic psychology among some Protestants'.[148] Yet at the same time, academic study of the latest census figures seemed to suggest that 'the chances of the Catholic community eventually outnumbering, and also out-voting, the Protestant majority are very slim ... it now looks increasingly likely that a stable balance somewhat short of the 50:50 mark will be the eventual outcome'.[149] The cry was taken up by the Unionist politician, John Taylor, who called upon Roman Catholics to realise that they would continue to be a minority. It may be that 'whatever the short-term opportunistic gains for either persuasion', the latest census figures available in the 1990s offered 'no real support for either stance'.[150] But for many Protestants the 'fear' was there, and could be dispelled not by statistics of emigration and population growth, but only by what one Presbyterian minister called 'a fundamental revision of the prevailing ideological construct of siege, isolation and defensive thinking'.[151]

7
Changing Times, 1960–1975

This is the day of protest and demonstration. Ban the bomb; end the war in Vietnam; stop South African rugby and cricket tours; aid Biafra; establish civil rights in Ulster. The placards are waved, the marchers chant their slogans, the TV crews work overtime and the world looks on as the student demonstrators lead the new protest movement. But protest provokes re-action. Demonstrators meet counter-demonstrators. Peaceful protest develops into violence. The police are involved and in the last resort, as in Ulster, the army is called in.

Herbert Carson, 1969[1]

New perspectives

'In retrospect, the 1960s, that time I remember as suffocating and dull', writes Fergal Keane, 'was a period of seismic change in Ireland.'[2] Mary Kenny 'remembered vividly' the revolutionary impact when she watched a BBC programme on the contraceptive pill in 1960.[3] Yet changes had already begun. People in the east of the republic had been able to tune into BBC television since 1953. In 1955 Ireland joined the United Nations, and in 1958 the IMF and the World Bank. In 1963 RTE began television broadcasts. Average annual emigration rates fell from 43,000 in the late 1950s to 16,000 in the early 1960s, and to 11,000 in the late 1960s. In 1963 Senator Ross assured pupils at Christ Church Cathedral School that they need no longer think of emigration, for although there had been a time 'when many Protestants felt that unless they could get work in a Protestant firm, they might not get work easily', now 'people no longer worried about whether one was a Protestant or not'.[4] Indeed as he spoke the Second Vatican Council in Rome was making decisions which would profoundly affect the way in which Roman Catholic and Protestants would inter-relate. That soon became clear in Ireland. When the Protestant benefactor of the Irish state, Chester Beattie, died in 1968, the funeral 'proved to be unique in Irish life as it was the first in a Protestant church attended by the Taoseach and President'.[5] But there was another side to the story. Archbishop McQuaid 'mistook the importance of the Second Vatican Council',[6] and indeed his own main contribution was to propose that the Virgin Mary be 'declared Mediatrix of All Graces'.[7] In 1966 he did not renew his ban on Roman Catholics attending TCD: was this evidence of the wind of change? That hope was dashed the next year, when the ban was reinstated, and when the Archbishop took to the columns of the press to reply to his critics.[8] In other ways too, things did not change. In 1962 Professor Enda McDonagh was telling the Maynooth Summer School 'that the superiority of virginity as a state in life was based on its exclusive choice of God, and it gave more intense expression to the mysteries of Christ'.[9] That piece of religious traditionalism was matched by a socio-political one, when in 1966 Dr T. O Raifeartaigh, Secretary of the Department of Education, lauded the men of 1916, 'the most noble generation the country ever produced', who believed that the Irish language should have 'its rightful place, that is the first place, in whatever education system would prevail in a free Ireland'.[10] In that year, the 50th anniversary of the 1916 rising was a reminder of much that divided Protestants and Catholics, southerners and northerners. But the wind of change was indeed coming in Ireland as it had in Africa. The *Irish Times* carried a buoyant editorial which in some ways was right, and in other ways was tragically wrong:[11]

> The bad old days, we said yesterday, are practically gone. Times are different in so many ways that we should not fear, whatever other troubles are in store for us, that religious faction fighting could be taken seriously, even in Belfast. The world has opened up: television, the all-round improvement in education, the speeding-up in communications of every kind, the disruption itself of the Second World War, all have broken in upon that narrow, bitter world in which the awful passions of religious warfare used to rage.

There were indications too that in the 'narrow, bitter world' of Northern Ireland, things were changing. In 1962 the IRA called off the sporadic campaign which had been launched in 1956. In 1963 Lord Brookeborough resigned after 20 years as Prime Minister. His successor, Captain Terence O'Neill, headed a government that seemed more flexible and forward looking. In June 1963 it sent a message of condolence to Cardinal Conway on the death of Pope John XXIII. In April 1964 O'Neill became the first Northern Irish Prime Minister to visit a Roman Catholic school. In January 1965 he entertained his southern counterpart, Seán Lemass, to lunch at Stormont, and a month later met him again in Dublin. Eddie McAteer led his Nationalists in Stormont to become the official opposition. Economic change was heralded by the publication in January 1965 of Professor Thomas Wilson's *Economic Development in Northern Ireland*, with its proposals for incentives and investment which were to pay off in the subsequent decade. The *Belfast Newsletter*, aware of the main thrust of the report, and of the new mood at Stormont, was hopeful that 'history may yet record 1964 as a turning point, as a year of greatness'.[12] Such hopes were evident elsewhere as well. In America John F. Kennedy in 1961 became the first president to have been born in the twentieth century. His brief and glittering presidency saw the growth of the civil rights movement, and his successor, L.B. Johnson, proclaimed the 'Great Society'. In Britain a new government took over in 1963. After 13 years of 'Tory misrule' Harold Wilson hailed a new Britain 'forged in the white heat' of the technological and scientific revolution. Already the 'Imperial Parliament', as it was still called in Northern Ireland, was undertaking an ever accelerating process of shedding its colonial possession in Africa and elsewhere. And the 'Swinging Sixties' would bring great changes in British society, not least in the realm of belief and religious practice.

Ecumenical activity was not the only religious development of the early 1960s. A whole spate of publications suggested that the shifts and developments in Christian thinking were as profound as those in politics and international affairs. Books like Alec Vidler's *Soundings* in 1962, and in 1963 Paul van Buren's *The Secular Meaning of the Gospel*, the collected essays on *Objections to Christian Belief* , and Bishop John Robinson's *Honest to God*, all seemed to suggest that 'our image of God must go', and that traditional beliefs must be re-interpreted in modern idiom. Of more immediate impact in Ireland was the attempt to re-present the Bible in terms that modern man could understand, with the publication in 1961 of the New English Bible, New Testament. It received a low-key welcome from the churches, Archbishop Simms commending it as 'a translation of the text which so accurately renders the original, while its diction is that of the present time, and the language is more readily understood by the younger generation of today'. Pointing out that it was being published on the 350th anniversary of 'our beloved Authorised Version', he noted that Synod was being asked to authorise its use in churches, which would 'in no way interfere with the status of the Authorised Bible'.[13] The Presbyterian welcome was even more circumspect, commending it 'for private study';[14] the Methodists were a little warmer: 'The Conference welcomes the *New English Bible (New Testament)* and commends the study of it to the Methodist people.'[15] New translations and versions appeared with increasing rapidity – the Jerusalem Bible (1966), the Good News Bible (1976) and the New International Version (1978) were particularly important, and cumulatively altered for ever the way in which Protestants (and Roman Catholics) perceived of, studied and quoted the Scriptures. There was resistance, and – as with Roman resistance to the disappearance of the Tridentine Rite, or Anglican dismay at the decreasing use of the Book of Common Prayer – some of the objections, perhaps most, were aesthetic and traditional. But there were also doctrinal objections. The Trinitarian Bible Society attacked the New English Bible for 'its theological bias and the inappropriateness of its language in many passages', as well as 'the quality of the underlying Greek text'.[16] The Revd Ian Paisley was more colourful in his objections, for he saw it as part of the ecumenical and modernist undermining of the faith. C.H. Dodd, the foremost scholar in the production of the NEB, was attacked as 'one of those Jesuit poisoned characters', who would naturally distort the Bible message: 'No wonder the *New English Bible* makes Peter out to be "the rock" on which the church was built.'[17] In the Articles of Faith of Paisley's Free Presbyterian Church, the 'Absolute and Divine Verbal Inspiration' of scripture refers to the Authorised Version only.[18] Paisley's objections to the NEB are symbolic of a wider reaction within Irish Protestantism against ecumenism and modernism.

The three main Protestant Churches had been relatively free from breakaway groups or internal divisions since at least the middle of the nineteenth century. The Church of Ireland, as we have seen, had occasional 'ritual' troubles, but no schisms resulted, and only a very few clergy left the church. The Presbyterian Church had a traumatic experience in 1927, when the brilliant young professor J.E. Davey of the Presbyterian College, Belfast was accused of heresy. He had been a fellow of King's College, Cambridge, and was appointed to Assembly's in 1918 at the age of 27. His books included *Our Faith in God* (1922) and *The Changing Vesture of the Faith* (1923) which were singled out for attack by his critics, who claimed he was denying the doctrines of the incarnation and the atonement. Tried before the Belfast Presbytery in 14 sessions in February and March 1927, he was acquitted. His case on appeal came before the General Assembly in June, and again he was acquitted, by 707 votes to 82.[19] The result was that a small group took themselves out of the church, and founded the Irish Evangelical (or from 1964 the Evangelical Presbyterian) Church, led by the Revd James Hunter of Knock, who 'regarded it as his supreme task to stem the tide of theological liberalism', and who had decided after the acquittal that 'the Church was less than loyal to her credal standards'.[20] Davey's biographer has noted the contribution of W.P. Nicholson to 'the turbulence in the atmosphere which surrounded the heresy trial', and 'the extravagant licence of his attacks upon "liberals", "modernists", "higher clergy" and Unitarians'.[21] John Barkley recalled that Davey 'was pilloried throughout the country in a public campaign of vilification'.[22] This was important, because the fall-out from the trial affected more than the Presbyterian church, and in a phenomenon which was also evident in the 1960s, created a populist mood of suspicion which infected even Presbyterians who did not chose to become schismatic. The breakaway church never became numerically significant; in 1995 it recorded a total communicant membership of 458, though with 13 churches and 10 full-time ministers. 'Subsequent events within PCI', its Clerk of Presbytery claimed in 1998, 'have fully vindicated the concerns of 1927'; these 'concerns' were: 'involvement in unbiblical ecumenism, an increased degree of liberal theology in Union Theological College and the ordination of women'.[23]

Ian Paisley was, as he proudly states, 'born the very year the Heresy Charges were laid against Professor Davey in 1926'. He was brought up in Ballymena, where his father had become pastor of Hill Street Baptist Church in 1928, and later formed a new 'Regular Baptist' Assembly in the town. After training at the Barry school of Evangelism, and the Reformed Presbyterian Church's Theological Hall in Belfast, Ian Paisley was ordained by his father in August 1946 as minister of the Ravenhill Evangelical Mission Church, established nine years previously as a breakaway from Ravenhill Presbyterian church. In April 1951 this church became the second 'Free Presbyterian Church of Ulster', the first having been inaugurated on St Patrick's Day 1951 in a mission hall in Crossgar. Paisley's interpretation of the origins of his church – that it split off from the Presbyterian Church rather in the same way that the Irish Evangelical church had done in 1927, and for similar reasons – has been substantially questioned.[24] But most certainly his new church appealed to disaffected Presbyterians, on the Ravenhill Road, in Crossgar, and at Drumreagh and Rasharkin, Co. Antrim, where the next two churches were planted. By 1966 there were 13 congregations. Apart from his fundamentalist Gospel preaching, his main draw card seems to have been his anti-Catholicism, well illustrated in the Maura Lyons affair of 1956. She was the 15-year-old daughter of a staunchly Catholic family from the Falls Road area of Belfast, who joined Paisley's church. Her home was then visited by three priests who, she later claimed, 'seemed determined to force me into convent life', so she fled and was spirited away to Scotland. In December Paisley staged a rally at the Ulster hall, when a tape recording of the girl's voice was played, recounting her story and expressing her fears. It was a theatrically triumphant version of the National Union of Protestants' tactic in having the ex-nun Monica Farrell speaking on their platforms describing the horrors of convent life, and an echo of the 'moral panic' induced by the McCann case in 1910. In the event, the girl returned in May 1957, and a court case ensued. She was placed in the care of her father, but her safety was guaranteed; in the event, she married and returned to the Catholic fold two years later.[25] But the publicity had been achieved. Paisley had played on some of the traditional fears of Ulster's Protestants – of the power of the Catholic church, the influence of its priests, and the deep suspicions of what went on in convents. There were no new perspectives here, but a rerun of old fears and hostilities, which were to become Paisley's stock in trade. It had its appeal. By 1981 Paisley's church had 49 centres in Ireland and 5 overseas. The 1981 census revealed it to be the seventh largest church in the Province. By then, Paisley was a political as well as a religious leader.

The Methodists faced their most significant breakaway in the wake of the events of the 1960s. The Irish Methodist Revival Movement was founded by some members of the church in Fermanagh and Tyrone in October 1964. In its first quarterly bulletin it spelt out its fundamental beliefs: the Bible as 'the inspired, infallible Word of God'; the 'twin doctrines' of justification by faith and scriptural holiness; the need for its members to 'have an experience of the new birth'.[26] The main thrust of its polemic borrowed much from the Paisley canon: against the 'Rome-ward trend' of ecumenism, as represented both in the World Council of Churches, and in the unity talks currently in train both in England (Anglican–Methodist) and in Ireland (Anglican–Methodist–Presbyterian). When the tripartite discussions led to the declaration of intent in January 1968, it was interpreted as 'the Pope's lieutenants ... uniting to form one Church, so that [the harlot Church of] Rome need not concern itself with the sordid and time-consuming business of separate denominational take-overs'.[27] Political issues impinged. The

Biafran civil war was 'a Roman Catholic civil war', because Ojukwu was a Catholic, supported by Portugal and by oil interests in France and Italy.[28] The Methodist College, Belfast, was attacked for allowing civil rights leader Rory McShane to address some pupils. The Revd Eric Gallagher was invited to 'take a walk around the Roman Catholic areas of Belfast and see the new housing estates and new schools. Where did the money come from for all these and many more things which have been given to a minority which has, for the past fifty years, been stabbing in the back those who have been feeding, housing and clothing them.'[29] And in its last issue, an initiative by the Ministry of Community Relations to bring together young Roman Catholics and Protestants in Londonderry was attacked as increasing 'the possibility of mixed marriages' and of 'another Protestant being won to the Roman Catholic system'. The same issue of the magazine attacked the New English Bible (the Old Testament had now been published), revealing 'numerous departures from the God-honouring, Christ exalting translation of the Authorised Version', and claiming that 'the discerning reader will be quick to see it as a counterfeit Bible acceptable only to a counterfeit church'.[30] Not surprisingly, the members of the movement soon after that decided to leave the 'counterfeit' church of which they were still ostensibly members. They initially found refuge in lining up with the American-based Free Methodist Church, and later in the creation of a Fellowship of Independent Methodist Churches. In 1998 it had 18 churches (2 in the Republic) and some 14 clergy. The Free Methodist Church by that time had 6 churches in the Province, and claimed a membership of some 200, while another Methodist-based denomination, the Church of the Nazarene, had 13 churches in Northern Ireland, with a membership of some 750.

These Methodist groupings are merely a small cross-section of the many new movements and churches which have grown up in Ireland – and particularly in Northern Ireland – over the course of the century. The Cooneyites or Dippers in Co. Fermanagh in the early years of the century,[31] and the Elim Pentecostalites who began in Monaghan in 1915,[32] are earlier examples. In the last quarter of the century, new fellowship and house groups have been a feature of the religious scene in Northern Ireland, often with charismatic features.[33] Inter-denominational, Province-wide movements like the Irish Alliance of Christian Workers' Unions, and the Faith Mission which reached Ireland from Scotland in 1890, have proclaimed the Gospel with fervour, and have often stimulated the creation of such groups or churches. The common characteristics of most of these groups pose a challenge to the larger churches. They offer the comfort and comradeship of small face-to-face societies, in the midst of increasing secularisation and anonymity, features of western society which were becoming more evident in Ireland in the late twentieth century. They tend to emphasise the 'Reformation Truths' – as they see them – in contrast to the perceived 'liberalism' of the established churches. They shun the ecumenical movement, and retain a huge distrust of the Roman Catholic Church. And they highlight a characteristic of the history of the Protestant churches described by Eric Gallagher and Stanley Worrall as 'Christian leaders making constant appeals for moderation and reconciliation, but forced always to look over their shoulders to see how far they were isolating themselves from their own flocks.'[34] *Whither Methodism?* abounds with references to 'the Irish Methodist hierarchy' or 'the "wise men" of the ecumenical field'. 'It has been untruly asserted', it complained, 'that the Irish Methodist Revival Movement is anti-clerical. In vain we have awaited a lead from some of our ministers.'[35] It is reminiscent of the attacks by the Cooneyites early in the century, who seemed 'to take a special delight in making frivolous charges of the most un-called for and uncharitable nature against ministers of the Gospel, forgetful that the apostolic circles and churches were not perfect'.[36] Although Edward Cooney in the early part of the century, or the leaders of the IMRM in the later part, may not have attracted large numbers of adherents, they undermined the leaders of the churches, forcing them to 'look over their shoulders'. A similar phenomenon in the Roman Catholic Church was noted in the last quarter of the century, when 'society's social and sexual mores had changed dramatically, leaving the relevance of the Church in question', the new questioning of authority 'heightened by resentment against Bishop Cahal Daly'.[37]

It is difficult to disentangle the effects of 'the Troubles' in Northern Ireland from the changes in perspective which were proceeding anyway from the 1960s onwards. One of those changes was in the role and position of women. Two signposts indicate the way things had been and were becoming. In the general election of 1964, there was only one woman candidate. Politics had almost always been, and still was, man's business, as would be expected in a fairly traditional and conservative society.[38] Just six years later, another signpost indicates how things were moving: the Northern Ireland government announced its acceptance of the principle of equal pay for women.[39] How did, and how would, the churches view the role of women? In the Church of Ireland, the structure of government set up at disestablishment excluded them at all levels; only men could serve as vestrymen, and thus become eligible to sit as members of a select vestry, a diocesan synod, or the General Synod. It was this all-male body of laymen and clergy who were confronted in 1914 by a petition from 1,400 women, and a motion proposed by J.A.F. Gregg and seconded by Mr Justice Madden, that parochial offices should be open to women, as they had been before 1870. It was defeated, with Bishop Bernard of Ossory vocal in opposition, even though he had (unsuccessfully) supported women's admission to TCD in 1892. (They were finally admitted there in 1903.) It was defeated again in 1919, but in 1920 was passed by 183 clerical votes to 27, and 84 lay votes to 43.[40] Nine years later, an attempt by Lord Glenavy to open up diocesan and General Synods to women was defeated by 58 clerical votes to 92, and 69 lay votes to 56. A 'Church of Ireland League' was formed to pursue the change,[41] and the motion was lost in 1930 by only one vote.[42] At the Down Diocesan Synod a few months later, Lord Cushendun spoke up for change, declaring that 'if the motion were for the purpose of admitting women to Holy Orders he could then very well imagine that there would be hesitation', but that by merely enfranchising women they would 'be removing a very real and disgraceful stigma'.[43] These enlightened opinions were not shared by most General Synodsmen, who voted the proposal down with increasing majorities in 1931 and 1932. The Presbyterians meanwhile had caught up. In 1926 it was agreed that women could become elders,[44] but four years later an attempt to allow women to become ministers was removed from the books, having found little support either among the Presbyteries or the members of the General Assembly.[45] Church of Ireland women had to wait until 1949, when their admission to all lay offices was overwhelmingly approved by 158 to 39.[46] In practice however, progress was slow thereafter with only one woman elected to the General Synod in 1951, and still only 18 (or some 7% of the lay total) in 1969. In 1972 the first women lay readers were instituted.[47] In 1973, the Presbyterians went further, and agreed to the ordination of women.[48] In 1974 the Methodists, following the practically unanimous vote of the British Conference in 1971, proposed that 'in view of our understanding of the mind of Christ, and the equality of the sexes, women who wish to offer themselves for the ministry of the Methodist Church in Ireland may do so under the same conditions as men'.[49] This resolution was adopted in 1975.[50] In 1978, just before the Lambeth Conference of that year, the Church of Ireland bishops declared that they saw no theological objection to the ordination of women. In 1968 a survey of a random selection of Irish Anglicans had revealed that 34% of the clergy and 43% of the laity were in favour of the ordination of women.[51] Ten years later, at the time of the Lambeth Conference, women had already been admitted to the Anglican priesthood in Canada, the USA, New Zealand and Hong Kong. It was not however until 1990 that the Irish General Synod passed the necessary legislation; 37 days after that, the first two women were ordained in St Anne's Cathedral, Belfast. There was little opposition; two senior clergy resigned.[52] By contrast, when the Church of England followed suit in 1992, hundreds of clergy defected, mostly to Rome. When Archbishop Carey visited the Pope four years later, he received a 'straight-from-the-shoulder rebuke', according to one Vatican official, for having taken this step.[53] Papal displeasure was unlikely to worry many Irish Anglicans. But the division of views it represented was significant. For Roman Catholics in Ireland, 'the church's image of the ideal woman was defined by Mary's role as wife and mother',[54] a model which had entirely fitted with de Valera's vision of Ireland. The new perspectives of the 1960s might offer different role models, when in Eire the women who 'seemed to excite public admiration most were Bernadette Devlin ... and the novelist Edna O'Brien'.[55] Traditional Catholic perceptions were rather curiously closer to the views of the followers of Paisley and the smaller churches, who similarly viewed women's role as a subordinate one; and they, too, were adamant in their belief in the Virgin Birth, a doctrine which they saw being challenged by more 'liberal' Protestants. The divisions which appeared in sharper focus from the 1960s onwards have meant, it has been suggested, that 'Ulster Protestantism is characterised not only by denominational cleavages but by a conservative/liberal divide.' A survey in 1983 delineated the nature of this divide: '76% of members of the Church of Ireland were "liberals", 53% of Presbyterians and 55% of Methodists, but only 6% of Baptists and 4% of Brethren'.[56] Although the terms 'conservative' and 'liberal' here connote religious or theological opinions, the divisions within denominations, and the reasons why leaders had to 'look over their shoulders', cannot be divorced from 'the Troubles' in Northern Ireland. These in turn were partly a result of the new perspectives in the Ireland of the 1960s.

The Northern Ireland problem

The 'seismic' changes – political and social – that were taking place in Ireland in the 1960s, and which mirrored significant changes throughout the western world, were matched by changes within the churches, and as we have seen, theological ferment and ecumenical dialogue were features of these years. 'Nearer my Pope to thee' was the way Ian Paisley greeted the meeting of Archbishop Fisher with Pope John XXIII in 1960, but very much more modest efforts at Protestant–Catholic reconciliation in Ulster were even more fiercely attacked. In 1964 the Revd Robert Nelson invited a Roman Catholic priest to speak at a meeting in his (Methodist) manse, and the Revd Jack Withers issued a similar invitation to a meeting at Fisherwick Presbyterian Church. The announcement that these meetings were to be picketed by Ian Paisley and his followers led to their cancellation, and to an (unsuccessful) attempt by the Armagh Presbytery to get a ban imposed on any such functions.[57] Towards the end of the year the 'threat of a return to the Troubled Twenties' increased, with the outbreak of three nights' rioting in the Divis Street area of the Falls Road, after Paisley had threatened to lead a group of his followers to remove a tricolour from the Sinn Fein Offices, unless official action were taken.[58] The RUC's removal of the flag, and its subsequent reappearance, sparked off the bloodiest strife for 30 years. Ex-Moderator Robert Corkey placed the blame at the door of Ian Paisley, who had 'won for himself a considerable following of thoughtless people, who are apparently as brainless as himself'.[59] It was an understandable reaction, but it – and other events of 1964 – illuminate much of what followed. Political developments (such as the Lemass–O'Neill meeting) and ecumenical developments (such the invitation to the priests) were seen as threatening to the Protestant religious and political ascendancy in Ulster; Paisley understood, articulated and exacerbated those fears, and was dismissed as an ignorant obscurantist. As further political and ecumenical developments unfolded, Paisley's popularity rose as he continued to give expression to the fears of many ordinary Protestants.

In 1962 Denis Barritt and Charles Carter had published their groundbreaking book, *The Northern Ireland Problem: A Study in Group Relations*. It was commissioned by the Irish Association, which had been founded in 1938 'to promote communication, understanding and co-operation between north and South, Unionist and nationalist, Protestant and Catholic'.[60] It drew attention, in an undramatic way, to instances of discrimination, though also to instances of exaggeration. It made some modest proposals for greater communication and more contact and understanding between Catholics and Protestants. 'If Northern Ireland does not adapt herself to a world grown impatient with petty disunity', the authors concluded prophetically, 'she will enter a period of increasing and painful stress.'[61] It was a book later commended by the Presbyterian Church for study by its members. In the wake of its publication, and of the events of 1964, including the founding of the Campaign for Social Justice, the General Assembly's Committee on

National and International problems decided to tackle the issue of discrimination. In 1964 its report had dealt with a relatively non-controversial issue, the care of the mentally ill.[62] In 1965 it tackled the 'new morality', in itself a reflection of the increasing concern felt among Christians about the 'permissive society' and the breakdown of traditional codes of conduct.[63] In that year too, it decided that religious discrimination would be its next area of study, and the General Assembly set the tone by passing a resolution urging 'our own people humbly and frankly to acknowledge and to ask forgiveness for any attitudes and actions towards our Roman Catholic fellow-countrymen which have been unworthy of our calling as followers of Jesus Christ' and asking them 'to resolve to deal with all conflicts of interests, loyalties or beliefs always in a spirit of charity rather than of suspicion and intolerance and in accordance with truth as set forth in the Scriptures'.[64] The report on religious discrimination was published in 1966, with a revised version in 1967 'commended for study to kirk sessions and study groups'.[65] It was a long and somewhat convoluted document.[66] It suffered from 'the initial failure to define religious discrimination', as one Presbyterian critic put it.[67] Its tackling of the franchise question was somewhat disingenuous; the report did not recognise a problem here. On the housing question, it was much more outspoken: 'We believe that the allocation of houses on the part of local authorities should be solely on grounds of need, and that anything in the nature of religious discrimination is wrong.' Overall the conclusion was that 'we realise that the elimination of religious discrimination is no easy task, but we welcome the signs that we are moving in the right direction'. The Methodist Church published a much shorter and much more outspoken statement on 'The Present Situation in Ireland', which included an abundance of scriptural texts, and spoke clearly against any attempts to prevent or interrupt lawful assemblies, any insulting of the faith of others, and all forms of injustice, inequality or discrimination based on creed, race or colour.[68]

All these discussions and reports took place against a background of considerable ferment. The Campaign for Social Justice was joined by other groups, such as the Campaign for Democracy in Ulster, founded in London in June 1965 with British Labour MPs among its constituency, the Northern Ireland Civil Rights Association (1967) and People's Democracy (1968). At the other end of the spectrum, protests continued. There was a controversy over the naming of the new bridge over the Lagan in Belfast. The somewhat hapless Edward Carson, son of Lord Carson, was imported by Paisley to raise the temperature, though

Carson's credibility with many must have been somewhat damaged by his remark on UTV that 'I don't want to see High Mass sung in St Anne's Cathedral.'[69] But in the year of the 50th anniversary of the Somme (and of the Easter Rising) it was a master stroke by Paisley to have Carson accompany him to the first rally of the newly formed Ulster Protestant Volunteers. He then organised a demonstration during the Presbyterian General Assembly meetings. He led his followers from Ravenhill Road, via the Markets where scuffles broke out between Catholics and the RUC. While those continued, Paisley and his followers proceeded to Howard Street, where they hurled abuse at the Presbyterian delegates, and their guests of honour, the Governor and Lady Erskine. For this affray, Paisley was imprisoned (having refused the option of a fine) and the 'martyrdom' became yet another weapon in his armoury. The Orange Order resolutions on the Twelfth of July that year were full of attacks on 'the present trend towards one united Church, involving the surrender of our distinctive Protestant witness',[70] and the Revd Donald Gillies (Agnes Street Presbyterian) praised them in the hope that they would 'lead many to alter course before we are wrecked on the rocks of ecumenism and driven ashore on Roman beaches'.[71] The build-up of opposition to the leadership of O'Neill on the one hand, and to any hint of 'appeasement' of Roman Catholics on the other, continued throughout 1967. Early in the year Dean Peacocke invited Bishop Moorman of Ripon to speak in St Anne's Cathedral. Moorman had been an Anglican observer at Vatican II, was chairman of the Joint Anglican-Roman Catholic Commission, and was also an expert on the history of the Franciscan Order. He was an obvious target. 'Evangelical Protestantism was enraged.'[72] The widespread outrage and threats of demonstrations were such that O'Neill privately asked Peacocke to cancel the invitation.[73] The Church, and the Prime Minister, succumbed to the pressure. 'I've often wondered', said a guest of Peacocke's at the dinner when O'Neill's request was made, 'what would have happened had we taken a firmer line with Paisley earlier on.'[74] Here was a dilemma which would confront church leaders again and again: was the line of least resistance the most eirenical response in the circumstances? Could church leaders risk taking a stand which might only increase the distance between them and their flocks, or drive some of their members out to the more extremist churches and groups? The Standing Committee of the Church of Ireland did make a statement regretting the threat to freedom of speech, but by then it was too late. Enraged Protestantism had issued its threats, and had won the day. The climb-down did not win the Church any friends. O'Neillite Unionism and ecclesiastical ecumenism came under further violent attack at the particularly unruly Twelfth of July demonstrations throughout the Province. The year ended with an unusual show of solidarity among the church leaders. Although they 'had not dared to meet together', they arranged – by telephone and emissary – to publish a joint new year appeal for peace, which appeared in the press signed by the Cardinal, Moderator, President and Primate.[75] They had undoubtedly read the results of a *Belfast Telegraph* opinion poll in December 1967: 35% of all unionists were reported to agree with Paisley's protests; 44% did not think he was trying to stir up sectarian bad feeling; 55% opposed the Unionist party curbing the influence of the Orange Order. Equally worrying, the poll revealed that Paisley's most significant support came from middle-aged and older Presbyterian workers in Belfast, people 'who felt most threatened by the changing climate around them and most nostalgic for the certitudes of traditional Unionism'.[76] Paisley's talent for playing the tunes to which such men would respond was nowhere better illustrated than in his patronage of a project to bring home the *Clydevalley* (or *Mountjoy II*), one of the ships used for the gunrunning into Larne during the great home rule crisis. It was refitted in Canada, and sailed into Larne in December 1968. Though only 4,000 instead of 25,000 turned out to welcome the ship, and although lack of funds resulted in her being broken up instead of becoming a museum, Paisley nevertheless managed to present himself as standing in the great tradition of Ulster's heroes in the fight for their civil and religious rights. Greeting the arrival of the *Clydevalley*, his newspaper printed a very long poem, the final verse of which gives the flavour.[77]

Men of Ulster! Here's your orders!
Take this mighty sword and fight;
This alone can give us victory,
Change the darkness into light:
Let us follow Christ, our Leader,
See! He points to yonder crown,
As we celebrate *Clyde Valley* [*sic*],
Sailing now for Larne town.

The Paisleyite political and religious rhetoric, with its populist anti-establishment twist and its evocations of Ulster's past, contrasts with what T.E. Utley called the 'charm and patrician hauteur' of Terence O'Neill, whose 'government by gesture' was by 1968 looking distinctly shaky. The gestures highlighted by Utley – 'dramatic speeches and appeals on television, for example, elaborate foreign tours, and overtures to the South and to the Catholic minority' – were counterproductive, because they 'excited Protestant suspicion without materially reducing Catholic grievances'.[78]

Those grievances enabled others, rather than the churches, to claim the moral high ground in the course of 1968. The Northern Ireland Civil Right Association (NICRA), founded in January 1967, had chosen its name well. Martin Luther King Jr had already achieved what would later be called 'iconic' status throughout the western world among the young, the disadvantaged, the liberal and the disaffected. He was assassinated in April 1968, and the Civil Rights Bill became law in the USA a week later. Against that background, the first NICRA protest march took place in August, and by its clever – and some would say misleading – representation of the grievances of Ulster Catholics as paralleling those of American blacks, it achieved world-wide publicity. Throughout 1968 and 1969 television was a powerful weapon in its armoury; it was easier for watchers to identify with those who sang the haunting 'We shall overcome' than with those who shouted the slogans of Protestant Ulster. It was all too easy for the world to see NICRA marchers as victims, especially when viewing the scenes of mayhem at the disruption of the Belfast–Derry march at Burntollet Bridge in January 1969. Any attempt by the churches to occupy the moral high ground was undermined by their divisions, within and between themselves. The Irish Council of Churches issued a statement in November emphasising the need for reforms in the areas of employment, housing allocation and local franchise, but also expressing the hope that Roman Catholics would 'seriously examine in what ways they can alleviate the effects of those of their policies which tend to divide the community'.[79] When a march was planned for Armagh in November 1968, the two primates jointly appealed for restraint; when it passed off relatively peacefully, the Cardinal praised the NICRA stewards, and the Primate, the RUC.[80] One of the Primate's own clergy – the Rector of Knocknamuckley, near Portadown – called upon him to petition the cardinal 'to prohibit the so-called civil rights movement marching through Protestant areas in this province of Ulster'.[81] Others of his clergy signed their names to a 'We back O'Neill' spread in the *Belfast Telegraph*.[82] When a civil rights march frightened the Protestant minority in Newry, Ian Paisley agreed not to bus in supporters, but the local RC Bishop took no action. After the march, with its attendant violence, the Protestant clergy spoke up on behalf of the genuine NICRA grievances, but attacked both the agitators who made use of such marches, and also the local RC clergy, who if they had shown a 'more noticeable and more active desire to preserve peace', might have helped the situation.[83] Ulster, as the Prime Minister had said in his pre-election broadcast in December 1968, was at the crossroads.[84] Could or should the churches point their people in a particular direction? The Presbyterian, John Withers, had complained that 'all over the world evil rages ... and the church keeps a dignified silence, refusing to take sides in the conflict'.[85] As Moderator he would not recommend to Presbyterians how they should vote, though he signed a significant statement – 'Church and Community' – which committed the church to 'Social justice and Christian charity', and, 'in a changing, revolutionary situation', to a ministry of reconciliation.[86] Twenty-two Presbyterian ministers issued a statement declaring that 'a wide cross-section of moderate and Christian opinion in the country' recognised 'the lasting benefits for the whole community under Captain O'Neill's leadership'. A hundred and twenty Anglicans produced a 'message of support from the Bishops and clergy' stating that 'the future well-being of this community can best be secured by the moderate reforming policies of the present Cabinet'. Archbishop McCann, who did not sign, was reported to be unavailable for comment.[87] People who would not normally speak out publicly on party political issues now did so. The Newtownabbey Methodist Circuit wrote to the Prime Minister 'wholeheartedly endorsing the spirit of your leadership, and its outworking in policies of moderation'. A Banbridge meeting of clergy sent a message of support to O'Neill, signed by three Roman Catholics, two Presbyterians, two Methodists, two Anglicans and a Baptist. The Churches' Industrial Council signalled its full support for the 'policies of reform, moderation and reconciliation emphasised by the Prime Minister in his televised speech'.[88]

The Stormont election in late February 1969 gave the Unionists an increased majority: of the 52 seats they won 39, as opposed to 37 previously. But they were divided between 27 pro-O'Neillites, 10 anti-O'Neillites, and two others whose allegiance was unclear. By the time the churches gathered for their respective annual meetings in June, O'Neill had resigned, replaced by his cousin, James Chichester-Clark, who defeated Brian Faulkner by one vote in the leadership contest. The printed record of the General Synod, meeting in May less than a month after O'Neill's resignation, contained no resolutions on the momentous events north of the border, and only the Primate's highly generalised call to 'stand firmly for the equal rights of every man irrespective of nationality, colour or religion', and to be 'seen to be firmly opposed to all forms of fanaticism – religious or political', give any indication of the grave concerns many members must have felt.[89] The Presbyterian General Assembly, the next month, also meeting on this occasion in Dublin, affirmed its support for the Northern government 'in its efforts to create a unified and prosperous community on the basis of equality, justice and reform', urged it to hasten in its 'elimination of all legitimate grievances', and paid tribute to 'the Royal Ulster Constabulary (without claiming that all are blameless) for the restraint, fidelity and courage shown in the performance of their duties in the face of sever provocation and bitter vilification'.[90] The Methodist Conference, also in Dublin that year, made what was the clearest and most direct statement on the events of the previous year, recognising that 'peace and understanding in Northern Ireland cannot be attained without full justice for all sections of the community', welcoming the reforms promised by government, and urging their speedy implementation. It urged 'all people, and especially Methodists, to make determined attempts to understand the hopes and fears of every section in the community'. It concluded by 'deploring acts and words of violence from whatever quarter', and in what was perhaps the only veiled criticism of other churches – particularly the Catholic – stated the hope that 'a joint call to reconciliation and full participation in the life of the community' might soon be possible.[91]

The two months following those church meetings in 1969 were turbulent, culminating in the new prime minister asking for British troops to help stem the rising tide of violence. They were deployed on the streets of Belfast and Londonderry on 14 August, where they found the clergy already active in the worst areas. Agnes Street and Shankhill Road Methodist churches in Belfast were kept open as 'a steady trickle of people – some terrified' arrived to pray for peace.[92] In Dungannon, two young Anglican curates walked the streets of the town throughout the night, pleading with people to go home, and issuing through the local newspaper a plea for peace.[93] In Belfast an Anglican cleric faced a 200-strong mob and successfully checked them by declaring, 'You pass this point over my dead body.'[94] Presbyterian and Anglican church halls were opened as help centres.[95] The value of the clergy was recognised in a government announcement that they did not require permits to enter and leave curfew zones.[96] In an 'incredible peace treaty', six members of Broadway Presbyterian church and six from St Paul's Roman Catholic parish set up a joint committee, which when he visited it the Archbishop of Armagh declared was a model for further such co-operation.[97]

That was almost the first duty of the new Anglican primate. George Simms had been born in Donegal, and educated in England, and at TCD.

He had taught there and at Lincoln Theological College, and had been Bishop of Cork and Archbishop of Dublin. He was not known in the north; his Anglo-Irish accent, his fluency in the Irish language, his gentle and somewhat elliptical style of speech were not attributes that would immediately endear him to Ulstermen, even of his own flock. 'It's not something anyone would seek', he had commented on hearing of his election to the primacy in July.[98] He must have felt that even more as southern Anglicans began to distance themselves from the events in the north. The Dean of Cork declared that 50 years of Unionist and Protestant government in the north had 'produced the present holocaust', whereas southern Protestants 'have been fairly and honourably treated as first class citizens and have been happily integrated into the community'.[99] Colonel and Mrs O'Callaghan-Westropp of Co. Clare inserted a special notice in a Dublin paper announcing that they no longer wished to be known as Protestants because 'we are so thoroughly ashamed of our co-religionists in the six counties'.[100] Even if these were emotional reactions, the official response from the south caused the northern Protestants to feel yet more beleaguered. The crisis was 'seriously exacerbated by an incredibly irresponsible intervention by the Dublin premier', writes Brian Faulkner.[101] Jack Lynch, stating that his government 'would not stand idly by', had dispatched army units and 'field hospitals' to the border areas, and a message to the UN calling for a peace-keeping force to intervene. To whom could Protestants turn? Their own church leadership lacked the continuity the Roman Catholics had, with their Moderator or President changing each year, and with a new Archbishop in Armagh. Indeed all the northern Irish bishoprics changed hands: Connor, like Armagh, in 1969, and Derry and Down in 1970.[102] R.C.P. Hanson, a Dubliner who had been a professor at Nottingham University, was like a fish out of water in the rural diocese of Clogher, and resigned, also in 1970, after an episcopate of just three years. Political personnel also changed. In 1970 Harold Wilson was replaced by Edward Heath. In 1971 Chichester-Clark was replaced by Brian Faulkner. In July 1970, a curfew was imposed on the Lower Falls area, and in a weekend of violence, the army fired 1,454 rounds, and used 1,600 canisters of CS gas. Five people in all were killed and some sixty injured. What this 'battle of the Lower Falls' did, as the provisional IRA leader Seán MacStiofáin later claimed, 'was to provide endless water for the Republican guerrilla fish to swim in'.[103] The result was soon to be the further escalation of violence in the Province. Meanwhile the church leaders were co-operating more closely. At the end of August 1969 Archbishop Simms, Cardinal Conway, and the

Presbyterian Moderator John Carson, were televised talking to Malcolm Muggeridge on BBC TV, and to Kenneth Harris on Ulster Television.[104] At its meeting in November the Irish Council of Churches decided that it had 'to face what form of relationship, if any, it should seek to have with the Roman Catholic church'.[105] Its discussions advanced rapidly enough for the churches in their mid-1970 meetings to endorse the proposal that joint working parties on social questions be set up involving the ICC and the Roman Catholic hierarchy. The Joint Group that emerged produced reports on drug abuse (1972), housing in Northern Ireland (1973), teenage drinking (1974) and underdevelopment in rural Ireland (1976). Also in 1976 it produced, as we shall see, its important report, *Violence in Ireland.*

The violence meanwhile continued, and the introduction of internment in August 1971 put a great strain on the churches' united stand. Cardinal Conway appealed for calm, but was soon expressing his community's abhorrence of internment, which had initially been used only to pick up Roman Catholics. The Protestant Church leaders were on holiday. Their appointed representatives, like the Cardinal, issued a statement calling for restraint, but came out 'in support of the Government's decision to intern, while regretting the necessity for its introduction'.[106] Archbishop Simms later spoke on RTE giving guarded support for the measure, while not revealing that he had been absent when the statement was drawn up, a fact revealed by the *Irish Times* early in September.[107] It was, say his friends, the one time they saw him really upset. But worse was to follow. The statistics of 'the Troubles' tell something of the story (Table 7.1).[108]

Table 7.1: Statistics of 'the Troubles', 1969–75

	Civilians killed	*Army/RUC/ UDR killed*	*Total killed*	*Shooting incidents*	*Bomb explosions*
1969	12	1	13	–	8
1970	32	2	25	213	153
1971	114	59	173	1,756	1,022
1972	322	145	467	10,628	1,382
1973	171	79	250	5,018	978
1974	166	50	216	3,206	685
1975	217	30	247	1,803	399

Some particular events stood out in the months after internment. The first priest to lose his life, Fr Hugh Mullen, was shot while administering the last rites. In December 15 Roman Catholics were killed in a bomb explosion at McGurk's Bar in Belfast. In January 1972 13 Derry Catholics were shot dead by paratroopers in what became known as 'Bloody Sunday', the reverberations from which would last into the new century. Three weeks later seven people were killed by an IRA bomb planted at Aldershot military barracks, and a few weeks after that the bombing of the Abercorn Restaurant in Belfast killed two people, and injured 130, mostly women and children out shopping. Three weeks after that horror, on Good Friday, the British government announced direct rule, the suspension of Stormont, and the end of half a century of devolved government in Northern Ireland. Few at the time would have believed that it would be a quarter of a century later, on another Good Friday, that an agreement on lasting peace and devolved government in the Province would be tentatively signed. Direct rule meanwhile seemed to make no impact on the spiral of violence. In July 1972 the Provisional IRA exploded 26 bombs in Belfast, killing 11 people and injuring 130. At the crowded Oxford Street bus station, where seven people were killed, the impact of the television pictures was enormous, as many people now viewed in colour the rescue workers tending the injured and gathering together the shattered remains of those who had been blown to pieces. It was dubbed 'Bloody Friday', partly in a Protestant response to the Catholic outrage over 'Bloody Sunday'.

The churches – Catholic and Protestant – were faced with an ongoing dilemma. Condemnations of violence went unheeded by the men of violence; failure to speak out on matters which profoundly affected a particular community led to a loss of influence by church leaders within that community. In the aftermath of internment, it began to seem to the Revd Eric Gallagher that there might be another way. He was probably the Protestant churchman most able and best placed to formulate another approach; he had been superintendent of Methodism's Belfast Central Mission for 15 years, and had been President of his church in 1967–8. He was well known in ecumenical circles, and had been responsible for many of the sane and balanced pronouncements which his church had issued over the years. In his own account he tells how, as he contemplated a possible personal approach to the IRA, he was contacted by Father Desmond Wilson, with a view to setting up just such a meeting.[109] Within a few weeks Gallagher was talking to Seán MacStiofáin, Rory O'Brady and Joe Cahill in the Imperial Hotel, in Dundalk. He expounded his faith, and his hopes for peace, and after an hour and a half, MacStiofáin suggested he pass a message to the British that a cease-fire might be possible. It was duly passed on. In June 1972 the IRA declared a two week cease-fire, in the wake of which the

Secretary of State, Willie Whitelaw, set up a secret meeting in London, at the house of one of his minsters, Paul Channon. The British government, as it had done so often in the past in colonial situations (and was soon to do again in Rhodesia), undertook talks with the 'men of violence'. The IRA team – which included Gerry Adams and Martin McGuinness – presented their demands. 'The meeting was a non-event', Whitelaw recalls.[110] MacStiofáin of course remembered things rather differently, and blamed the subsequent breakdown of the truce on British perfidy, leading him to announce that the campaign of violence would 'be resumed with the utmost ferocity and ruthlessness'.[111] As the figures given above show, that is what happened. With the Tartan Gangs and the UDA on the loyalist side becoming better armed, there was talk of a Protestant backlash, if not all-out civil war.

Talks with the IRA might not lead to a settlement, but talks with the constitutional parties might, and for a brief moment in 1973–4 the 'Sunningdale Experiment' held out the hope of a power-sharing between nationalists and unionists, between Catholics and Protestants, though its all-Ireland dimension frightened and angered many unionists. That fear, and the fall of the Heath government which had devised it, and the Ulster Workers' Council strike of May 1974, killed off Sunningdale. But the precedent for direct talks with the IRA had been set, and churchmen then took up the challenge. Through channels available to the Revd William Arlow, associate secretary of the ICC, an invitation was received from the IRA, the British government was informed, and a meeting took place in December 1974 in a hotel in the village of Feakle, Co. Clare. The Protestant participants in addition to Arlow, were Bishop Butler of Connor; Harry Morton and Arthur McArthur of the British Council of Churches; Jack Weir, Clerk of the Presbyterian General Assembly; Ralph Baxter, secretary of the ICC; Eric Gallagher, and Stanley Worrall, a leading Methodist layman and ex-headmaster of Methodist College, Belfast. They presented the IRA with a draft which they thought might let them 'off the hook' of violence without loss of face. Full discussion was not possible because a raid by the Irish police meant that most of the IRA men had to leave hurriedly, and it was the police who broke to the world the news of the meeting. There was no substantial agreement by the IRA leaders to the churchmen's draft. But a cease-fire was announced, and the British government responded by a lower level of army activity. It held over Christmas. Never formally revoked, it gradually broke down, and even more people were to die in 1975 than in 1974. It was small comfort to the churchmen that internment, which by 1975 was being criticised by the ICC as well as by the Roman Catholic bishops, was abandoned towards the end of that year. That removed at least one element of what it had become fashionable to call 'institutionalised violence'. But terrorist violence continued. The Constitutional Convention of 1976 – another attempt at 'power sharing' – was even less successful than Sunningdale. The words of the Feakle churchmen, or of the politicians, seemed less significant than the bombs and bullets of the men of violence.

8
Towards the New Millennium

> In Northern Ireland both communities have hopes concerning the future of the province. For some their hope is that we will continue to be part of the United Kingdom, for others it is that we will become part of a United Ireland. Some in each community desire first and foremost that reconciliation and peace will come. For those of us who claim to live according to Christian standards or consider ourselves to be Christians we can and must submit our hopes and how we live under their influence, to God's authority and examine them in relations to the principles He provides for us in the Bible.
>
> *A Christian Response to the Irish Situation, 1997*[1]

Violence in Ireland

'Ulster has reached its millennium of murder and misery', the *Belfast Telegraph* declared in 1974, printing the names of all 1,000 people who had been killed since the beginning of 'the Troubles'.[2] In 1980, two more 'grim milestones' were passed: the 2,000th death, and the 100th death of a member of the Ulster Defence Regiment.[3] In 1991 the 2,000th civilian to be killed marked another 'grim landmark', with a total of 2,911 deaths in all.[4] Had the same proportion of people been killed in Britain, the total would have been over 100,000. That was the scale of the problem of violence. Every so often there appeared to be defining moments, bringing hope that this might be a turning point. Such a moment came in August 1976, when three young children were killed in Andersonstown, and gave rise to the Peace Movement led by Betty

Williams and Mairead Corrigan. Another was the Enniskillen bomb in 1987, during the Remembrance Day service, and the death of a young nurse whose father, Gordon Wilson, emerged as a foremost spokesman for peace. The Omagh bombing in August 1998 was another. Every death of course brought grief to a family, and sometimes the poignancy was unbearable for many, especially when children were the victims – the young son of a Presbyterian minister, two small boys in Warrington, three small boys in Ballymena and many more. And every murder, especially of a civilian, was followed by what became an 'almost ritualistic condemnation session at the BBC'.[5] The murders were ghastly, but so too for many was the world in which they happened. 'Is there life *before* death?', asked a graffito in Ballymurphy in 1973.[6]

The Inter-church Group which was set up in 1970 produced its fifth report in 1976 – *Violence in Ireland*. The joint chairmen of the group responsible were Bishop Cahal Daly and the Revd Eric Gallagher. These two were among the outstanding churchmen of the century. Daly had 'masterminded Ireland's implementation of the revolutionary theological perspectives of the second Vatican Council'.[7] Many thought he should have succeeded Conway as Archbishop of Armagh in 1976, but that promotion, and a Cardinalate, came to him only in 1990. To many people he seemed to be 'Irish Catholicism's one great prophetic leader of this generation'.[8] Eric Gallagher, as we have seen, was the leading Methodist, possibly the leading Protestant churchman, of his generation. These two chaired a group which included Denis Barritt whose seminal study of the Northern Ireland problem had appeared in 1962, Stanley Worrall, who had attended the Feakle talks, and David Bleakley, prominent Anglican layman and sometime Minister of Community Relations. Despite the profoundly different traditions from which the participants came, and despite the fact that the Northern Ireland problem had been characterised world-wide as a religious conflict, the unanimity which the participants achieved was remarkable. The first section, on 'Violence in Ireland', was an historical survey which traced the roots of discord in Ireland from the distant past through to the civil rights movement and the subsequent IRA campaign. That an agreed statement on the history of violence could be produced did something to show that, while communities need stories about their own past in order to understand themselves, those stories can in a sense be shared and balanced. The second section, on the results of violence, was inherently less controversial. Quite apart from the physical and economic consequences, the report raised questions about the possible long-term moral and psychological effects of 'the Troubles' on the people of the Province, especially its young people. The third section set out the teaching of the churches, showing that it was possible for Protestants and Catholics, and even for pacifists and non-pacifists, to reach an agreed statement on Christian teaching on the state, war and personal behaviour. The fourth section, on 'The Task of the Churches', presented the group with the most difficult problems. They could point to increased co-operation between the churches, and the proliferation of groups like Corrymeela and the Assisi Fellowship, which worked for peace and reconciliation. More difficult to identify was a programme for future action by the churches. Much of this section involved an emphasis on 'justice': that churches and their members must act justly, must uncover injustice and come to the aid of its victims, and must seek to oppose injustice, by non-violent means. They noted 'two live issues which have caused misunderstanding between the Churches and concerning which a closer accord would contribute to the cause of reconciliation'[9] – 'mixed marriages' and 'separated education'.

Archbishop Simms had chaired the Anglican–Roman Catholic International Commission on the Theology of Marriage, which met between 1967 and 1975. Pope Paul VI's 1970 motu proprio, *Matrimonia Mixta*, reflected the new ecumenical mood, in that the Protestant partner in a mixed marriage was no longer required to give a promise regarding any children of the marriage. The final report of the Commission went further: as well as recognising the validity of any such marriage before a minister of either church, with no necessity for a declaration by the non-Catholic partner, it also endorsed the dropping of the promise by the Catholic partner as well, in favour of an assurance by the priest to his bishop that the partner had been apprised of his or her obligations. While the 1978 Lambeth Conference welcomed the report,[10] it was a 'great disappointment' to Simms that it was never formally adopted, and he felt that as a result of the 'snail-like progress of the churches', many couples simply drifted away from church participation altogether.[11] In the early years after the publication of *Matrimonia Mixta*, Church of Ireland clergy in particular were still unhappy about the situation. The Dean of St Patrick's claimed in 1972 that mixed marriage problems were of much more concern as a source of tension in the Republic than were issues like the Constitution and censorship.[12] In the same year, Bishop Armstrong of Cashel (later of Armagh) told his flock that there 'was little real change in the position regarding mixed marriages',[13] while the Anglican newspaper declared that 'the spirit of *Ne temere* survives in Ireland despite Cardinal Conway's recent assurances'.[14] Others were less gloomy. A Methodist minister in Dublin in the same year felt that the

Catholic Church's attitude had 'changed very greatly'.[15] An examination of Quaker mixed marriages concluded that between 1971 and 1990 'the loss of Friends who married Catholics has equalled the gain of former Catholics who chose to join the Society'.[16] The Northern Ireland Mixed Marriages Association was founded in 1973.[17] The number of mixed marriages does not seem to have increased dramatically, but by the 1980s attitudes to them, and to much else, had become more tolerant.

Perhaps there was something of a predisposition in the *Violence in Ireland* report to interpret the milieu in which the churches were operating as being socially static. It had looked that way. The 'permissive society' took time to reach Ireland, and by the mid-1970s the changes that were evident elsewhere in western societies had still to impact significantly on both parts of the island. In 1972 the publication of the Longford report on pornography elicited small response in Belfast where a street poll indicated that everyone interviewed, from housewives to students, agreed that there was no pornography 'problem' locally, and the following year an English journalist could but agree, as he searched the bookshops in vain.[18] That same journalist noted that 'they banned the film "Last Tango in Paris", showing once again that Belfast isn't all bombs, there's a nice bit of censorship too'. But the next month saw an announcement that there would be 'no more silly controversies about the censorship of films', as the 'Belfast City Council seems set on allowing the British Board of Film Censors' certificate to be the standard of taste for the public here.'[19] Yet when *The Life of Brian* (a Monty Python spoof on the life of Christ) appeared seven years later, its distributors withdrew it when the Belfast City Council insisted on an 18+ certificate; the same happened, it must be said, in Dublin and Glasgow.[20] While the perception of a widespread acceptance of 'permissiveness' in Britain may have owed much to the power and influence of the London based press and television, there were nevertheless enormous shifts in both attitudes and practices in the last quarter of the century. They were evident, with a time delay and sometimes only partially, in Ireland. A survey in 1990 is revealing. When asked whether pornography should be banned altogether, 42% of men in Northern Ireland and 31% of men in Britain said yes. For females the responses were 66% and 45% respectively. But among the 18–24 age group, the difference between males in Northern Ireland and Britain was less marked: 22% as opposed to 19%.[21]

While it might appear that 'the strength of Irish Presbyterianism arose, in part, from its cultural isolation',[22] there was also an understanding that the church had to serve and witness 'in a time of convulsive upheaval',[23] and that it was 'well-nigh impossible for the processes of change and the spirit of protest to escape the notice of any society'.[24] But the Committee on the State of Religion in 1971 reported – rather despairingly – that of the 364 sessional returns, 220 churches stated that 'the social and political upheaval of recent years' had created no special problems. Was this, the Committee asked, based on 'a realistic assessment' or 'born of fatal complacency'?[25] Quite apart from 'the Troubles', other problems were beginning to have an impact. At a Police Surgeons conference in Dublin in 1970, the rise in drug taking was noted. 'The first groups of young drugtakers were seen in Dublin in 1967', mostly belonging to the lowest social class; by 1970 it was evidently a problem affecting 'children from so-called respectable homes'.[26] Archbishop Simms highlighted the problem at the 1971 General Synod, and praised 'a group of our younger clergy' who were tackling it. He reported that the Inter-church Joint Group would be looking at this issue as its first task.[27] Another challenge to complacency was the issue of homosexual rights, and the time-lag between Northern Ireland and Britain is well illustrated here. The Province was not included in the 1967 Act that permitted homosexual acts between consenting adults in private. The Gay Liberation Movement did not reach Ulster until 1973, when branches opened at the two universities.[28] In 1974 two Irish delegates attended the first international Gay Rights Conference in Edinburgh, from which an open letter to the Christian Churches emerged. Transmitted to the churches in Ireland by the Irish Gay Rights Movement, it elicited no response from the Roman Catholic Church. The Methodist President, Harold Sloan, 'replied immediately with a concerned and helpful letter', and Archbishop Buchanan of Dublin, replied similarly.[29] The 1976 General Synod voted in favour of decriminalising homosexual relations between consenting adult males. The Methodist Conference of 1977 did the same, adding that the church had a role 'in the education of society ... so that ignorance, prejudice and fear may disappear with increased knowledge and understanding'.[30] In what many homosexuals saw as a 'sop to the hard-line evangelical lobby',[31] the Methodist declaration, like that of the Anglicans, included a statement that it was 'opposed to all debased forms of homosexuality and heterosexuality, eg., prostitution, promiscuity and assault upon minors'. In 1982, Northern Ireland came into line with the rest of the United Kingdom, despite Ian Paisley's 'Save Ulster from Sodomy' campaign. It was another 11 years before legislation in the Republic was passed to the same end. The crisis later in the century over homosexuality among the clergy did not, however, become an issue in Ireland, while it threatened to dominate Anglican debate in England in the early

1990s. In 1992 Archbishop Eames declared 'that there were no practising homosexuals among the clergy, and that he could not see how an active homosexual could become a priest in the Church of Ireland'.[32] The Roman Catholic Church, which had opposed the changes in the law both north and south, fared much worse, and a 'spate of sex scandals' began to look like 'gravely damaging that disturbed and floundering vessel' in the early 1990s.[33]

That attitudes were changing in the whole area of human relationships was becoming evident in the 1980s. During the heated debates about decriminalising homosexuality, a 1977 Presbyterian call for 'sympathy, compassion and understanding for homosexuals', was tempered by a feeling that the 'mood of the age is such that we are being asked to accept everything as natural'.[34] Here, albeit in moderate tones, was the voice of those who feared where it would all stop. Many felt that way, and did not go unheard. In that same year, a new TV sex education series, *Man and Woman*, containing 'plain facts' about sexual relations, was screened throughout the United Kingdom – except in Northern Ireland.[35] But when an Inter-church Group examined 'Marriage and the Family' in 1987, it was in no doubt that during 'the 1970s there was an erosion of traditional authority which, in some places, has been dramatic. The influence and authority over young people by traditional structures such as the family and the Churches have been weakened.'[36] Statistics bore this out. A survey in the Republic showed that from the mid-1970s to the mid-1980s the percentage of Roman Catholics who believed that premarital sex was always wrong had fallen from 71% to 46%. Divorce petitions filed in Northern Ireland had risen from 123 in 1961, to 435 in 1971, to 1,673 in 1981, the law having changed in 1978. In an Anglican parish in west Belfast in 1984–5, 30% of baptised children came from single-parent families. Illegitimate births were 2.5% of the total in Northern Ireland in 1961, and 7.8% in 1982. Even more dramatic changes were reported in the Republic, where births out of wedlock increased from 1.6% in 1960 to 19.5% in 1993.[37] In the light of these changes, the concern expressed in *Violence in Ireland* about mixed marriages looks less significant. It was estimated in 1995 that they accounted for between 3% and 6% of all marriages contracted in the Province. Set against the statistics for divorce or single-parent families, it was hardly of great significance, and that is just one illustration of the way things had changed since 1974.

The other issue raised by *Violence in Ireland* was 'separated education'. 'Members of our Working Party are not in agreement on this issue', the Report unsurprisingly declared.[38] As we have seen, education was a deeply divisive issue. Throughout the western world the 1960s were a time of student activism in universities, and a consumer revolution was in train which was partly a cause of, and partly the result of, a huge new teenage market. Young people aged 18 and over now had the vote. Schools, like universities, were looked at afresh by a society alive to the new-found spending power of young people, and by a government increasingly aware of the national need for an educated workforce. Independent schools were under threat from a Labour government which regarded them as perpetuating the class system. Northern Ireland's traditional education system, and the problems inherited from before partition and reinforced since, would begin to look increasingly out of tune with the spirit of the times. As independent schools in England might be helping to create and reinforce the class system, so Northern Ireland's schools might be helping to create and reinforce the sectarian divisions in society. At the Unionist Party's annual conference in 1969 – in that confused period between the 'crossroads' election and the resignation of Terence O'Neill – segregated education came under attack, described by one delegate as 'the most blatant piece of religious discrimination in the province at present' – presumably because of the subsidising of Roman Catholic Schools. What looked like a 'liberal' motion – but which was more a veiled attack on a minority seen as increasingly vociferous and demanding – called for integrated education in all schools, and was carried by 278 to 239, though opposed by the Minister of Education, Captain Long.[39] Two years later the Ulster Teachers Union unanimously passed a resolution that 'integrated education is essential to an integrated community in Ulster'.[40] Lord MacDermott, at the first annual MacDermott Lecture at Queen's University in 1972, declared that 'the time has come to reconsider, on a basis of tolerance and consent, the opening of all our schools to the children of all religions'.[41] The Archbishop of Canterbury, Michael Ramsey, preached in St Anne's cathedral in 1974 and spoke in favour of integrated education. The Catholic Bishop of Down and Connor protested. 'There is no question, as you suggest', Ramsey replied to the Bishop, 'of my encouraging Catholic parents to violate their consciences.' It was the tragedy of the huge sectarian divide, he wrote, 'which causes me, and others to hope that it may be possible to have a school system in which children while being taught the faith of their own Church are not segregated from the children of other Churches'.[42] That possibility was already being actively considered in political circles. Lord Wolfenden was commissioned in 1973 by the J. Rowntree Memorial Trust to study and report on the education system in Northern

Ireland. It was never published. Basil McIvor, the education minister in the short-lived 'Sunningdale' power-sharing executive, did not know why it was suppressed, but 'wondered had it recommended some form of integrated schooling'.[43] McIvor had become convinced that shared schooling was the way forward, and opinion polls among both Catholic and Protestant parents seemed to suggest that the time was ripe. In late April 1974, just a few weeks before the five-month power-sharing experiment collapsed, McIvor proposed to his colleagues the creation of '"shared schools", available to Catholic and Protestant parents alike who wished to see their children educated together'. This might avoid a head-on clash with the Roman Catholic Church, while responding to a groundswell of opinion both inside and outside the Province. However just before presenting his proposal, he recalls that he 'received a telephone message from Cardinal Conway warning me not to interfere with the schools'.[44] He went ahead and got the support of his colleagues. Despite the fact that 'some of the SDLP members had reservations about the details of the scheme', Brian Faulkner recalled, 'no one wavered from collective responsibility, even when the cardinal and the Catholic bishops launched a strong attack on the whole idea'.[45] There was an encouraging initial response 'from the three main Protestant Churches and two of the teachers' unions',[46] but 'the hopes of dealing with this important idea, like many other hopes for Ulster, died with the executive'.[47] Nevertheless the hope that a 'third way' in education might found was not dead. Already some Roman Catholic parents were sending their children to state primary schools, though their actions drew criticism from both ends of the spectrum. The Catholic Church criticised them because, as one parish priest put it, 'it showed an appalling lack of appreciation of one's faith',[48] and there were reports that such children would have the sacrament of confirmation withheld from them – which was an exaggeration with just enough truth in it to be worrying. Some Protestants on the other hand complained that a Roman Catholic priest might join the rota of clergy teaching religious education, and thus Protestant children might be exposed to Catholic influence.[49] A group of Catholic parents formed ACT – 'All Children Together' – to 'find ways of providing religious education for Catholic children outside the Catholic school system'.[50] It was at this point that *Violence in Ireland* appeared. Confining themselves to matters on which they were agreed, the members of the working party did not tackle the question of shared or integrated schools. Rather they proposed areas in which co-operation might take place: more meetings between Catholic and Protestant children for cultural and sporting activities; teacher exchanges; possible shared sixth form colleges or nursery schools; an ecumenical dimension to any religious instruction in schools. Co-operation, rather than structural changes, was an agreed area which would indeed bear fruit in the future. 'Had this model', as one writer commented in 1995, 'been got under way then, in 1976, in however modest a manner, it might now have been bearing fruit in practical and crucially important ways.'[51] But the idea of structural change – the creation of 'shared schools' – was still being canvassed.

Basil McIvor's plans had not resulted in the necessary legislation, for power had returned to Westminster after the fall of the power-sharing executive. In 1978 a bill in the House of Lords, sponsored by Lord Dunleath, a Northern Ireland Alliance Party peer, created a mechanism whereby an 'integrated' school could be set up, with joint Protestant and Catholic control. In 1981 ACT decided to move ahead with the opening of such a school – to be called Lagan College. It had, and continued to have, a mix of some 52% Protestant children and 48% Catholic. Within ten years it catered for 740 pupils and 47 staff, and had gone through various changes of status. In 1984 it became a 'voluntary aided maintained' school, with 85% government capital funding. In 1990 it took advantage of new legislation to become a 'grant maintained' integrated school, with 100% government capital funding. It was an experiment greeted more enthusiastically outside than inside the Province. The Roman Catholic Church's attitude was clear; that of the Protestant churches less so. In 1990 Basil McIvor, chairman of the Lagan board of governors, received the Templeton Award on behalf of the College, and complained that the churches had been 'hanging back' from supporting integrated education.[52] The truth was more complex. In the wake of his proposals in 1974, the Presbyterians had passed a resolution 'reaffirming their belief in the principle of integration of day schools and recognising the benefit to community development that would ensue'.[53] The Anglicans were much more cautious, claiming that 'any prospects of integration can only be seen within the context of relations within the community', though they did suggest, in ways that prefigured the 1976 *Violence* report, various means for 'the encouragement of harmony and good relations within the community'. They were also aware that the church might be accused of inconsistency, given its 'different attitudes when we are in a Protestant majority situation, or a Roman Catholic majority situation', though 'the alternative to our own church schools in the Republic is somebody else's church schools'.[54] The Methodists too were cautious, 'convinced that the schooling of Roman Catholic and Protestant children together is but one step toward improved relationships'.[55] Opposition on the part of the Catholic Church, and reservations on the part of the Protestant churches, were reflected in political attitudes then and later. A survey of political groups in 1990 revealed that only the Alliance party 'approved' of integrated education, while the SDLP, Sinn Fein and the Ulster Unionists had 'reservations'.[56]

Two initiatives which did however received the backing of the churches were 'Education for Mutual Understanding' (EMU) and the common Religious Education syllabus. These followed, if belatedly, the directions suggested in the 1976 Report, and were a response to the Education Reform (Northern Ireland) Order of 1989. Launched in 1989 for implementation in 1990, EMU was 'a commitment to the promotion of mutual understanding among young people in Northern Ireland yet without threatening particular cultures and traditions'.[57] This was to be one theme in a cross-curricular agenda including cultural heritage, economic awareness, information technology, and careers and health education. Northern Ireland's citizens of the twenty-first century were to be educated out of at least some of the ghetto mentalities and narrow horizons of the past. The reaction of one Dungannon sixth former was, however, understandable: 'It is not us, but our parents and teachers who need EMU.'[58] The other initiative was a joint syllabus for religious education, drawn up in 1990, partly to ensure that it did not become marginalised in the new legally required common curriculum. Ian Paisley attacked the minister, Dr Brian Mawhinney (himself a Protestant Ulsterman), for sanctioning EMU and then inflicting on children an RE syllabus, as 'the only diet that can be fed to them with the blessing of ecumenism and Rome'.[59] In fact he was not alone in feeling disquiet, which had spread – though for different reasons to his – throughout the churches. Such unease crossed the sectarian divide, both in terms of the need for ensuring a place for religious education, but also with regard to the state's seeming preference for integrated education, and the consequent marginalisation of the churches. In 1990 the Roman Catholic Church clashed with government over inequities in funding for their schools as opposed to new integrated schools, and by an 'historic compromise' ensured that 100% funding would henceforth be available, although it forfeited the legal right to an entrenched majority on the governing bodies of such maintained schools.[60] The Protestant churches, so long used to state co-operation in the field of education, were concerned about the need 'to protect the Protestant ethos of controlled, state schools'.[61] Both the Catholic and Protestant churches were aware, as the millennium came to an end, that they faced threats to their traditional concepts of education. Economic considerations and secular agendas posed problems no longer rooted in the old sectarian divide. There was perhaps a need for 'the main-stream churches together ... re-examining their fundamental attitudes to schooling in Northern Ireland'.[62] EMU and the common Religious Education syllabus had already illustrated that they could act together to achieve ends which they desired in common. Perhaps the churches in Ireland would find that in the twenty-first century those things that they held in common were more important than issues which divided them.

Sectarianism and secularism

The Protestant churches, and especially the Church of Ireland, struggled to maintain their all-Ireland identity. The problems involved were highlighted at a 'mini-Lambeth with Celtic overtones', held in Dublin in 1982. The Anglican bishops from Scotland and Wales joined their Irish brethren and found much in common, particularly in the problems they faced, All three churches were bilingual; all had run-down inner-city parishes (Belfast, Glasgow, Cardiff); all had to maintain a witness in sparsely populated areas (Connacht, the Highlands, Snowdonia). All emphasised their 'consciousness of a continuity in their church life which goes back before the 6th century', and they all wished 'there were some other term than "Anglican" to describe the world-wide Anglican Communion'.[63] These problems were not new, nor were they confined to the Celtic fringe. Fourteen years earlier the Church of Ireland had looked seriously at its whole structure, to determine what 'changes, reorganisation and reforms' might be necessary 'if the Church is to have its full impact in modern society'.[64] Comparisons then with the Church in Wales had been instructive (Table 8.1).[65]

Table 8.1: Comparisons between the Church of Ireland and the Church in Wales, 1967

	Total Church of Ireland	*Total Church in Wales*
Dioceses	14	6
Incumbencies	648	854
Clergy	796	984
Deans	29	6
Archdeacons	27	14
Canons	145	65
Balance sheet assets	£17.5m	£16.4m

The Church of Ireland looked exceedingly top-heavy. *Administration 1967* proposed a reorganisation of the dioceses, leaving the church with the same number in all, but creating a new diocese of Belfast. The number of dignitaries would be pruned, though not drastically. The size of the General Synod would be reduced from 216 clerical representatives to 167, and from 432 lay to 334. Its most radical proposals concerned clerical tenure, the aim being to make the clergy more mobile, and appointments more flexible. That, more than anything, provoked the opposition which led to the report being ditched. Yet its spirit and recommendations were very much a product of the time when it was produced. The Church of England had just recently considered the Paul Report,[66] which made similar recommendations regarding clerical tenure, and which – though flawed in its assumptions about future developments[67] – was animated by a similar concern, well summed up in *Administration 1967*: 'Only if the Church is more interested in what it is going to become than in what it has been, will it seek continuous renewal.'[68] The Presbyterians also found that challenge difficult, though not in terms of structure and organisation: as a church concentrated mainly in the north east it had an inherently more cost-effective and efficient manpower structure than the Anglicans. But, more than the Anglicans, they suffered from a sense of confusion about what their church was 'going to become'. The effectiveness of the Ballymascanlon talks was being queried in 1977 by Church of Ireland clergy,[69] but the Presbyterians were more fundamentally worried. In 1974, an unsuccessful attempt was made in the General Assembly to preclude further involvement, and resolutions on the tripartite talks between the Protestant churches received only the most lukewarm support.[70] As we have seen, in 1978 they suspended their membership of the WCC, and withdrew in 1980.[71] In 1982 'deep dissatisfaction' was expressed regarding the Ballymascanlon talks, though a move to withdraw from them was again defeated.[72] Six years later, however, the Presbyterians withdrew from the tripartite talks. As one Presbyterian plaintively put it: 'If we cannot agree with Protestants, how can we expect to agree with Roman Catholics?'[73] The papal visit of 1979, the hunger strikes of 1981, and the Anglo-Irish Agreement of 1985 all served to destabilise the Protestants generally, and the Presbyterians in particular. Although only the Presbyterian Moderator among the main church leaders refused to be involved with the papal visit, even the future bishop of Cork recorded that 'as I watched and listened to the events at Knock, I felt in a very deep manner (as I had scarcely ever felt before) that I was a stranger in a strange land – "in" but not "of" the community. It was a strange feeling of isolation, compounded by the phrase "Mary, Queen of Ireland".'[74] The new Catholic Primate, Tommy Fee, had adopted an Irish version of his name – Tomas O Fiaich – when he took office in 1977, and in 1979 he became a cardinal.[75] Although outspoken in his condemnations of violence, his stance regarding the hunger strikers in the Maze prison caused Protestants concern. 'One could hardly allow an animal to remain in such conditions, let alone a human being', the Cardinal declared in 1978,[76] and was attacked by a former Presbyterian Moderator[77] and by the Rector of Hollywood who claimed that his remarks 'could have very serious results in widening the gulf between the two communities'.[78] That gulf was further widened by the Anglo-Irish Agreement signed at Hillsborough by the British and Irish Prime Ministers in November 1985. It created 'a climate of fear and hysteria in the Protestant Community'.[79] Just as George Simms had arrived in Armagh to face almost immediately the crisis over internment, so now Robin Eames was enthroned in February 1986, in a Province where, as he himself noted, '[M]oderation within the unionist family was swept aside.' It was the Protestants' turn to feel 'the first pangs of the alienation which, for other reasons and in other ways, their nationalist neighbours had complained of for years'.[80] The Anglican and Presbyterian churches condemned the agreement; the Methodists may have felt the same, but took no vote.[81] Other, non-political, aspects of the 'Thatcher years' may well have been of greater long-term significance. Zygmunt Bauman has spoken of the 'urge of mobility, built into the structure of contemporary life [which] prevents the arousal of strong affections for places; the places we occupy are no more than temporary stations'. There was a 'precarious belonging' built into modern life. 'Skills, jobs, occupations, residences, marriage partners – they all come and go, and tend to annoy or bore or embarrass when they stay too long.'[82] Among even the Baptists of Belfast it was noted that they had 'lost their zeal for evangelism so evident in earlier years', because of 'increasing affluence, mobility and the spirit of the age'.[83] The 'secularisation' which Britain had experienced was belatedly hitting Ireland. It could still be claimed in 1995 that 'Northern Ireland is a confessional society [with] one of the highest records of church attendance in Europe, for example, second only to that of the Republic of Ireland.'[84] An ICC survey in that year saw 'some signs of erosion', but came broadly to the same conclusion.[85] 'Secularism', as Archbishop Eames pointed out at the opening of the 1991 General Synod, might be 'an opportunity, not a defeat, for the Christian ethos'.[86] 'The process of secularisation', a Roman Catholic priest noted, was at work on both sides of the border, and might for his own church 'prove a more redoubtable opponent than anything it has faced previously'.[87] Writing in the context of ECTI (Evangelicals and Catholics Together in Ireland) Bishop Miller of Down identified secularism, consumerism and postmodern scepticism as forces redrawing the sectarian map of Ireland. The fault lines were no longer between Catholics and Protestants, but 'between those believe and those who no longer have any faith, and between those who espouse a traditional revealed Trinitarian faith and those who feel free to rewrite the tenets and morals of the Christian faith'.[88] Might all this cut off the oxygen to the growth of sectarianism? That was still, according to the Anglican Primate, 'the greatest challenge to our joint church action and the greatest problem for the Churches in Ireland, North and South',[89] speaking after the publication of a joint working party report on sectarianism.[90] There was a danger though that the church leaders, and middle-class church members, were rather distant from the grass-roots sectarianism of the Falls Road or the Shankill. The Methodist Church warned its members: 'Perhaps John Hewitt's verdict could be applied to us when he said "You coasted along" as he accused the comfortably off who tried to pretend to themselves that bigotry was evaporating from their vicinity.'[91] The Archdeacon of Dublin took up the same theme later in the decade, after a month of sectarian killings in Ulster: 'Sitting in our ecumenical and ecclesiastical armchairs, we may roundly condemn such atrocities before returning to our usually polite and leisurely conversations about inter-communion. Where is the active leadership at the senior level in any of our churches that indicates any awareness that unresolved tribal and denominational prejudices filter down to street level in forms that give credibility to men and women possessed by hatred and fear, and allow them – even encourage them – to feel justified in committing atrocities?'[92] The WCC had been a target for such criticism in the past, from Presbyterians, and also from Anglicans.[93] One Presbyterian noted that 'that there is a real danger of the Councils of Churches becoming structures on their own account – led by personnel who are less and less accountable to the membership of the Churches they represent. The WCC indeed is often castigated by its Protestant critics as rapidly becoming a super-Church, but divorced from the realities of Christian life and witness at the "grass roots".'[94] Such criticism could also be directed against the 1990s 'fashion' for apologising. The Archbishop of Canterbury preached in Dublin in 1994, in the presence of the Irish President, and asked 'forgiveness for our often brutal domination and crass insensitivity in the 800 years of history of our relationships'.[95] Cardinal Daly, in Canterbury Cathedral the following year, asked the English to forgive 'the wrongs inflicted by Irish people'.[96] When the British Prime Minister was reported as 'apologising' for the famine, a Presbyterian General Assembly member queried the 'rightness or usefulness of government (or Church) representatives taking on themselves the burden of issuing apologies for national misdeeds committed long ago'.[97] Such apologies seemed very remote from the Orangemen camped at Drumcree Church, or Catholic youths suffering punishment beatings in Belfast.

'Protestant and Catholic churches', as the *Irish Times* pointed out in 1990, 'have been highly successful in creating community', or rather, 'mutually exclusive communities'.[98] The problem with 'the "warmth" of community' within Ireland, claimed another observer, was its exclusivity, for while successfully translated from rural to urban settings, it could not embrace those across the sectarian divide, let alone groups like travellers, Jews and blacks, who 'have threatened the very nature of "community" in Ireland'.[99] The community in which most Irish Protestants lived towards the end of the century was hideously disfigured by decades of internal violence. Fergal Keane, reporting on the murder of Tutsis in Rwanda in 1994, recalled 'Seamus Heaney's line about "Each neighourly murder" in the backroads of Fermanagh'. He thought of Northern Ireland, where he 'had reported on numerous cases of people being murdered by men who worked with them or who bought cattle and land from them'.[100] The smallness, even intimacy, of the world within which the Irish Protestant churches operate has led outsiders to see it as a 'place apart'. It was claimed that 'Northern Ireland has already become as "separate" from both the UK and the rest of Ireland as England and Scotland are from the Republic.'[101] An 'Ulster nationalism' seemed to be reflected in the widespread substitution of the Union Jack by the Ulster flag, the red cross of St Patrick on a white background, with the red hand of Ulster, surmounted (or, now more often, not surmounted) by the crown. William Craig had told a Vanguard rally in 1972: 'If we are not going to get a fair deal inside the United Kingdom, we will, as our fathers and grandfathers did in 1912, take whatever steps are necessary',[102] and in 1973 the Vanguard Assembly member, Glenn Barr, made what has been called the first 'Ulster nationalist' speech at Stormont when he declared: 'I have no intention of remaining a British citizen at any price.'[103] It was a theme which reappeared regularly throughout the 1970s.[104] In the early 1980s James Callaghan was advocating 'a broadly independent state' for Northern Ireland,[105] though the Orange Order was sceptical about the feasibility of an independent Ulster, 'politically or economically'.[106] It was not an issue confined to Northern Ireland, as the new Labour government in the 1990s devolved power from London to Edinburgh and Cardiff. In a 'home rule all round' strategy reminiscent of Liberal thinking a century before, a similar policy was to be applied in Ireland. That 'giant step', as Archbishop Eames called it,[107] was represented by the Good Friday Agreement of 1998, when the majority of people in Ireland, north and south, voted for a devolved government in Northern Ireland, with an 'Irish dimension'. As the century ended, it was still unclear whether 'lasting peace' was possible on that basis. And within the churches themselves there were many non-political issues which would grow or diminish in the new millennium. Would Catholics and Evangelicals make further common cause against the forces of secularisation and doubt? Would 'Celtic Spirituality' become a distinctive feature of Irish Christianity, and retain its popularity beyond the shores of Ireland, or prove to be a transient response to the contemporary demand for 'greenness' and for stress relief? What of the many communities and projects that had sprung up, largely but not exclusively as a result of 'the Troubles'? Did places like Corrymeela or the Christian Renewal Centre at Rostrevor, or newer ventures like Forthspring, or the many cross-denominational encounter groups, herald a shift (not exclusive to Ireland) away from the previous focus on the institutional churches? Indeed would the Irish continue to retain high levels of church attendance in the third millennium?

Northern Ireland, as George Boyce has pointed out, 'stands not as a place apart, but rather at a confluence of various, and at times competing, influences and cultures'.[108] Many of the strands are religious: historic Presbyterian radicalism, historic Anglican Orangeism, high contemporary rates of church going, a conservative moralism, a streak of religious fundamentalism. These blend with an attachment to the land, and an affection 'based on a profound feeling for the region's lakes and mountains, rivers and valleys, the grand seascapes, and the splendid variety of flora and fauna'.[109] They blend too with a neighbourliness, and indeed an 'Irishness', exemplified in the 'crack', the gossip and chat beguiling to visitors even in war-torn Belfast.[110] But would the new world of the internet, along with television and cheap foreign travel, change all that by exposing the Irish to more and more outside influences? But even if northerners and southerners, Protestants and Catholics, shared a certain 'Irishness', as the twentieth century ended, it looked as unlikely as ever that they would share a common political system, even though they had a common and increasingly deep involvement in the European Community, and in the world. 'The politics of aerial photography would make an interesting study', writes Geoffrey Wheatcroft; 'An aerial photograph says that Portugal and Spain are one country, so are Norway and Sweden ... The thing about aerial photographs is that they don't show actual people, or their political wishes, or their sense of national identity.'[111] Viewed from the air, Ireland is one green island. And although its people might end the twentieth century still divided by a political border, its churches remained, as they had been, institutions covering the whole of Ireland. Irish Christians might not know what changes and chances they would face in the new millennium, but they could perhaps face them echoing the words of the Presbyterian Moderator a century earlier:[112]

> I can never let go the hope that, were God's Spirit poured out on us abundantly, and all prejudice and party spirit sunk, the representatives of our great Churches which God has blessed about equally in their efforts on the mission field, would find that they agree more than they thought they did, and would by friendly consultation come to see things so far eye to eye as to discover some method by which our divisions might be removed.

Notes

See the Bibliography for full bibliographical details of publications cited here. Official Church records are cited fully in the Notes, but the following abbreviations should be noted:

MCM	*Irish Methodist Conference Minutes*
MCR	*Irish Methodist Conference Reports*
JGS	*Journal of the General Synod of the Church of Ireland*
GAM	*Presbyterian General Assembly Minutes*
GAR	*Presbyterian General Assembly Reports*

Introduction

1. Daly, 'A Vision of Ecumenism in Ireland', p. 15.
2. Berger, 'Against the Current', p. 32.
3. Bowen, *History and the Shaping of Irish Protestantism*, p. 459.
4. See Munson, *The Nonconformists*, pp. 191–6.
5. McCaughey, *Memory and Redemption*, p. 26.
6. Akenson, *God's Peoples*, p. 354.
7. Davis, 'Nelson Mandela's Irish problem', p. 50.
8. See my *Humphrey Gibbs*, pp. 126, 156.
9. Dunlop, *A Precarious Belonging*, p. 144.
10. Daly, *op. cit.*, p. 23.
11. This is the theme in Clayton, *Enemies and Passing Friends*.

Chapter 1: The churches in 1900 – an anatomy

1. Quoted in Bullock, *After Sixty Years*, p. 125.
2. Colley, *Britain: Forging the Nation*, p. 369.
3. Bell, *Randall Davidson*, p. 232; Davidson was Bishop of Rochester (Canterbury from 1903).
4. Noll, *A History of Christianity in the United States and Canada*, p. 311.
5. *Witness*, 17 August 1900.
6. *Free Church Year Book and Official Report ... 1900*, p. 17.
7. Stewart, *The Narrow Ground*, p. 113.
8. Booth, founder of the Salvation Army, published his *In Darkest England, and a Way Out* in 1890.
9. John Wesley's revised Prayer Book, used by many Methodists in Ireland, also excluded these items.
10. Seddall, *The Church of Ireland*, p. 248.
11. Bell, *Disestablishment in Ireland and Wales*, pp. 195–6.
12. Rutherford, *Christian Reunion in Ireland*, p. 67.

13. Bernard, *The Present Position of the Irish Church*, p. 30.
14. *Irish Churchman*, 18 December 1914.
15. Ervine, *Sir Edward Carson*, p. 33.
16. *Irish Churchman* (17 January 1929) was still tussling with the issue, claiming that in 1885 the Attorney General and Solicitor General had given legal weight to the title.
17. D'Arcy, *Adventures of a Bishop*, p. 29; he mispells Midleton.
18. Figgis and Drury, *Rathmines School Roll*, p. 164.
19. Acheson, *A History of the Church of Ireland*, p. 256; as he ruefully says, 'It remains to be seen what effect the regrettable decision to end boarding at Portora from 1995 will have on the episcopate of the future.'
20. Clerical salaries are listed in the *Irish Church Directory*; see also McDowell, *The Church of Ireland*, p. 74.
21. Armstrong, *My Life in Connaught*, p. 265.
22. Birmingham, *Pleasant Places*, pp. 88–9.
23. *Church of Ireland Gazette*, 12 May 1961.
24. *The Warden*, 7 October 1910.
25. *Witness*, 22 June 1900.
26. Moody and Beckett, *Queen's, Belfast*, vol II, p. 493.
27. Figures given in Allen, *The Presbyterian College, Belfast*.
28. Hamilton, *History of Presbyterianism in Ireland*, p. 33.
29. Barkley, *Blackmouth and Dissenter*, pp. 13 and 178.
30. Dunlop, *A Precarious Belonging*, p. 144.
31. *Witness*, 11 March 1881.
32. Moody, *Memories and Musings*, p. 212, written in 1938.
33. *Witness*, 24 January 1902.
34. A Methodist describing the General Assembly: *Methodist Newsletter*, June 1980.
35. Dunlop, *op. cit.*, p. 8.
36. Moody, *op. cit.*, p. 15.
37. *Irish Presbyterian*, March 1902.
38. *Witness*, 2 Jan 1903.
39. Brown, 'Life after Death?', p. 62.
40. *The Warden*, 11 March 1910; leading English nonconformists like Spurgeon and Parker earned £1,500 and £1,700 respectively: Munson, *The Nonconformists*, p. 109.
41. Macaulay, 'The Price of our Presbyterianism', p. 6.
42. Ministers could opt to continue to receive the state subsidy during their life-time; almost every minister however 'commuted' – that is agreed that the money should be vested in a church trust fund.
43. Allen, *op. cit.*, Belfast, p. 299.
44. Cooke, 'Church Methodists in Ireland', p. 140.
45. Jeffery, 'Church Methodists in Ireland', p. 75.
46. Rigg, *A Comparative View of Church Organisation*, p. 207.
47. *Christian Advocate*, 21 March 1890.
48. *MCM, 1905*, pp. 108 and 123.
49. Cole, *History of Methodism in Ireland*, p. 171.
50. Harte, *The Road I Have Travelled*, p. 99.
51. McCrea, *Irish Methodism in the Twentieth Century*, p. 28.

52. Quoted in Smiley, *The Life and Letters of the Revd W. Smiley*, p. 79.
53. Troeltsch, *The Social Teachings of the Christian Churches*, vol. 2, p. 933.
54. See Peter Embly, 'The Early Development of the Plymouth Brethren' in Wilson, *Patterns of Sectarianism*, ch 7.
55. Harbinson, *No Surrender*, p. 54.
56. *Belfast Telegraph*, 13 November 1971: article on the 'Hallalujah Days in the Lough Road [mission] Hall' in Antrim in the 1920s.
57. *Historical Sketch of the First Church of Christ, Scientist, Belfast.*
58. Thompson, 'The Origin of the Irish Baptist Foreign Mission', p. 19.
59. Dowling, 'A History of the Irish Baptist College', pp. 29–37.
60. Report contained in John Baxter, 'Some Recollections of Cliftonville Moravian Church and Sunday School 1909–24': PRONI T/3582/2.
61. *Ibid*, p. 4.
62. Ervine, *Reminiscences*, p. 159.
63. See especially Hatton, *The Largest Amount of Good.*
64. *British Friend*, 6th month, 1899, p. 137.
65. *Ibid*, 1901, p. 138.
66. *Ibid*, 1899, p. 137.
67. Grubb, *Quakers in Ireland*, p. 126.
68. Grubb, *The Grubbs of Tipperary*, p. 141.

Chapter 2: The churches and politics, 1900–1922

1. *Belfast Newsletter*, 20 January 1912.
2. Mackail and Wyndham, *Life and Letters of George Wyndham*, vol 2, p. 436.
3. Quoted in Ervine, *Craigavon: Ulsterman*, p. 111.
4. McCarthy, *Priests and People in Ireland*, p. 7.
5. *Ibid*, pp. 278–9.
6. *Ibid*, p. 280.
7. *Ibid*, pp. 541–60.
8. Larkin, 'Church, State and Nation', p. 1276.
9. Rolleston to Hannay, 12 March 1905; in Rolleston, *Portrait of an Irishman*, p. 52; he did add that the one thing not absorbed was the Gaelic League.
10. *Irish Ecclesiastical Record*, vol IX (January–June 1901), p. 262.
11. *Ibid*, p. 374.
12. See Sybil Baker, 'Orange and Green: Belfast 1832–1912', p. 800.
13. McDowell and Webb, *Trinity College, Dublin*, p. 367.
14. Quoted in Patterson, 'Independent Orangeism', p. 8.
15. *Witness*, 3 January 1902.
16. Jackson, 'Irish Unionism and the Russellite Threat', p. 403.
17. Corkey, *The McCann Mixed Marriage Case*, pp. 14–15; see also his autobiography, *Glad Did I Live*; also Lee, 'Intermarriage, Conflict and Social Control in Ireland'.
18. J.B. Armour to his son, 9 February 1911: in McMinn, *Against the Tide*, p. 87.
19. Harte, *The Road I Have Travelled*, p. 92.
20. Holmes and Urquart, *Coming into the Light*, p. 98.
21. Lee, *op. cit.*, p. 17; 'Moral panics', he suggests, 'can be thought of as collective expressions of moral indignation.'

22. *Witness*, 14 January 1910.
23. *The Warden*, 21 January 1910.
24. Report dated 4 January 1911: in Buckland, *Irish Unionism*, p. 303.
25. Letter of 29 July 1911: in Ervine, *op. cit.*, p. 185.
26. Stewart, *The Ulster Crisis*, p. 48.
27. *Ulster Echo*, 21 December 1911; the papal directive or 'motu proprio' was usually referred to in Ulster as 'Motu Proprio' (the name of the type of directive) rather than by its actual title, *Quantavis diligentia*.
28. Walker, 'The Irish Presbyterian Anti-Home Rule Convention of 1912', p. 74.
29. Buckland, *Irish Unionism*, pp. 78–9.
30. *Evening Dispatch*, 23 May 1893.
31. Walker, *op. cit.*, pp. 75–6.
32. Buckland, *op. cit.*, p. 79.
33. Barron, *The God of My Life*, p. 105.
34. *Ibid*, p. 209.
35. *Witness*, 16 August 1912.
36. Barron, *op. cit.*, pp. 212–16.
37. Belfast Presbytery Minutes, 18 July 1912: PHS.
38. Buckland, *op. cit.*, p. 80; see also full report in *Christian Advocate*, 15 March 1912.
39. Diary entry by George Riddell, 14 April 1912; in *The Riddell Diaries*, ed. McEwen, p. 40.
40. Blake, *The Unknown Prime Minister*, p. 130.
41. *JGS, Special Meeting*, pp. xlvi–lii.
42. *Irish Churchman*, 26 April 1912; *Church of Ireland Gazette*, 12 July 1912.
43. *Witness*, 28 June 1912.
44. *Belfast Newsletter*, 8 July 1912.
45. *Ibid*, 13 July 1912.
46. *Irish Churchman*, 30 August 1912.
47. Minutes of Ulster Day Committee, 27 August 1912: PRONI D 1327/3/1.
48. Peacocke to Bernard, 1 September 1912: BL Add MS 52782.
49. Ervine, *op. cit.*, p. 235.
50. Moody, *Memories and Musings*, p. 12.
51. *Fermanagh Times*, 22 August 1912.
52. *The Times*, 6 May 1913.
53. *Irish Churchman*, 10 October 1913.
54. Hammond, 'The Religious Question in Ireland'.
55. Sykes, *Man as Churchman*, p. 164.
56. Conroy, *Occasional Sermons, Addresses and Essays*, p. 133.
57. *Belfast Newsletter*, 3 December 1889; Balfour was speaking in Glasgow.
58. *Hansard*, 3rd series, vol xlv (21 January 1897), cols 257–62.
59. *Ibid*, cols 267–8.
60. *Belfast Newsletter*, 8 June 1897.
61. *Freeman's Journal*, 9 June 1897.
62. *Hansard*, 4th series, vol l (9 July 1897), col 1532.
63. 'The Irish University Question', p. 105.

64. 'Memorandum on the Irish University Question, written by Mr Haldane at the request of A.J.B.', 20 October 1989: Cabinet Papers, vol xlviii, PRO CAB 37/48, 77.
65. PRO CAB 37/48, 82 (12 November 1898).
66. Salmon to Carson, 21 February 1899: BL Add MS 49709, 100.
67. *GAM, 1900*, p. 1000; *1901*, pp. 85–6; *1902*, pp. 289–90.
68. *MCM, 1900*, p. 88; *1901*, p. 92; *1902*, p. 104.
69. *First Report of the Royal Commission on University Education in Ireland*, p. 13.
70. *Ibid*, pp. 61ff.
71. *Ibid*, p. 188.
72. McDermott to Bryce, 22 March 1906: Bryce Papers, NLI MS 11012.
73. Bryce, 'University Education in Ireland', 17 December 1906: PRO CAB, vol lxxxv, CAB 37/85,99.
74. *Belfast Newsletter*, 26 January 1907.
75. Birrell, *Things Past Redress*, p. 67.
76. Birrell, 'University Education in Ireland', 19 November 1907: PRO CAB, vol xc: CAB 37/90,99.
77. *Hansard*, 4th series, vol clxxxvii (31 March 1908), col 397.
78. McDowell and Webb, *op. cit.*, p. 367.
79. Lord Roberts to Major-General H.H. Wilson, 24 March 1914: in Beckett, *The Army and the Curragh Incident*, p. 225.
80. *Ibid*, p. 29.
81. Corkey, *Glad Did I Live*, p. 116.
82. Dooley, 'The Organisation of Unionist Opposition to Home Rule', p. 60.
83. Buckland, *op. cit.*, p. 250.
84. King, *Ulster's Refusal to Submit*, p. 21.
85. King and McMahon, *Hope and History*, p. 35.
86. Smith, *James Nicholson Richardson*, p. 166.
87. Good, *Ulster and Ireland*, p. 145.
88. Ferguson, *The Pity of War*, p. 209.
89. *Ibid*, pp. 209–10.
90. *Irish Churchman*, 2 October 1914.
91. Bernard, *In War Time*, p. 5.
92. Fitzpatrick, in Bartlett and Jeffery, *A Military History of Ireland*, p. 388.
93. Jackson, 'Unionist Politics and Protestant Society in Edwardian Ireland', p. 860.
94. Fitzpatrick, 'The Logic of Collective Sacrifice', p. 1029.
95. Bell, *A History of Scouting in Northern Ireland*, pp. 19 and 45.
96. Johnson, in Kennedy and Ollerenshaw, *An Economic History of Ulster*, p. 184.
97. Hannay to Bernard, May 1917, in Murray, *Archbishop Bernard*, p. 303.
98. Bernard, *op. cit.*, pp. 105–6.
99. O'Farrell, 'Millenialism, Messianism and Utopianism in Irish History', p. 57.
100. Newsinger, 'I bring not peace but a sword', p. 615.
101. Irish Guild of the Church, Minute Book: RCB 131.
102. Milne, *Protestant Aid*, p. 17.
103. *GAM, 1916*, p. 100.
104. Murray, *op. cit.*, p. 281.

105. Letter of 5 May, quoted in McDowell and Webb, *op. cit.*: the authors comment 'it seemed rather strange that an Archbishop should be the man to express [this opinion] so forcibly', but in the circumstances, perhaps not.
106. Sibbett, *For Christ and Crown*, pp. 300–1.
107. Kennedy, *The Widening Gulf*, p. 27.
108. Moody, *op. cit.*, p. 37.
109. Orr, *The Road to the Somme*, pp. 221–2.
110. Bowen, *History and the Shaping of Irish Protestantism*, p. 380.
111. *GAM, 1918*, p. 564.
112. Bernard's Memorandum on his and Viscount Midleton's meeting with Lloyd George, 5 December 1917: BL Add MS 52781.
113. Irvine, *Northern Ireland: Faith and Faction*, p. 206.
114. Irish Guild of the Church Minute Book, 14 May 1918: RCB 131.
115. *Ibid*, 23 January 1920.
116. Note on the back of a letter dated 16 July 1917: BL Add MS 52782.
117. See Butler, 'Select Documents XLV: Lord Oranmore's Journal', p. 560.
118. Midleton to Bernard, 16 December 1921: BL Add MS 52781.
119. Kenny, *Goodbye to Catholic Ireland*, p. 101.
120. See, for example, Hastings, *A History of English Christianity*, p. 29.
121. *Belfast Newsletter*, 14 August 1919, in Kennedy, *op. cit.*, p. 31.
122. Bowen, *op. cit.*, p. 387.
123. Boyce, *Englishmen and Irish Troubles*, p. 47 and *passim.*
124. Kennedy, *op. cit.*, p. 39, and *passim* for other incidents.
125. Dooley, 'Monaghan Protestants in a Time of Crisis', p. 238.
126. Kenny, *op. cit.*, p. 109.
127. *GAM, 1922*, p. 103.
128. *JGS, 1922*, pp. liii–lv.
129. *Ibid, 1923*, p. 81.
130. Bence-Jones, *Twilight of the Ascendancy*, p. 209.
131. Dillon, *God and the Gun*, p. 168.
132. Quoted in Hennessey, *A History of Northern Ireland*, p. 31.
133. *GAR, 1923*, p. 21.
134. *Ibid., p.* 3.

Chapter 3: The gospel and society, 1900–1914

1. Clow, *Christ in the Social Order*, p. 254.
2. Mearns, *The Bitter Cry of Outcast London.*
3. *In Darkest England and a Way Out* was published in 1890.
4. Currie, *Methodism Divided*, p. 178.
5. Chadwick, *The Victorian Church, Part I*, p. 325.
6. His episcopal address in *Down Diocesan Synod Report*, 1910.
7. Magee, *A Short Account of the History ... of the Assembly's Mission in Dublin.*
8. Sibbett, *For Christ and Crown*, p. 168.
9. Irish Temperance League Minute Book, 16 December 1881; 24 March 1899.
10. *Ibid*, 23 October 1885; 14 November 1890; 2 December 1898.
11. *Ibid*, 4 December 1896.
12. *Christian Advocate*, 26 January 1894.

13. *MCM, 1884*, p. 78.
14. *GAM, 1891*, p. 73.
15. *Ibid*, p. 157.
16. Mathias, 'The Brewing Industry, Temperance and Politics', p. 108.
17. Malcolm, *'Ireland Sober, Ireland Free'*, p. 331.
18. Pyper, *The Irish Sacramental Wine Association*, p. 2.
19. *Irish Ecclesiastical Gazette*, 4 February 1888.
20. *GAM, 1875*, p. 806; *1888*, p. 544.
21. McCartney, *Nor Principalities Nor Powers*, p. 257.
22. Davey, *A Memoir of the Revd Charles Davey*, p. 69.
23. Jeffery, *Irish Methodism*, p. 69.
24. Burns, *Temperance in the Victorian Age*, p. 106.
25. See, for example, Irish Temperance League Minute Book, 14 May 1886.
26. See *Towards Reconciliation*, pp. 77–8.
27. *Irish Temperance League Journal*, 1 January 1893.
28. *GAM, 1904*, pp. 812–15.
29. Morgan, 'A Study of the Work of American Revivalists', p. 465.
30. *Ibid*, p. 444.
31. *MCM, 1886*, p. 82.
32. Contemporary estimate quoted in Sibbett, *op. cit.*, p. 212; also in *Christian Advocate*, 8 February 1889.
33. Belfast Presbytery Minutes, Report of Denominational Statistics Committee, 2 February 1892: PHS.
34. The Revd W. Alford, *Fermanagh Times*, 30 March 1893.
35. Newman, *Change and the Catholic Church*, p. 150.
36. Nicholas, *Christianity and Socialism*, pp. 132–3.
37. *Ibid*, p. 166.
38. *Christian Advocate*, 18 August 1893.
39. Quoted in *Ibid*, 15 September 1893.
40. Norman, *Church and Society in England*, p. 144.
41. Minute Book of the Junior Ministers' Convention, 1898: WHS.
42. *MCM, 1908*, p. 121.
43. Dublin Presbytery Minute Book, 8 October 1901: PHS.
44. *GAR, 1904*, pp. 812–15.
45. *JGS, 1904*, p. 290.
46. Pooler, 'The Socialist Movement', p. 302.
47. See *GAM*, *MCM*, and the Diocesan Minute Books of Down, and Dublin, for these years.
48. See, for example, *MCM, 1912*, p. 118; *GAM, 1914*, p. 865.
49. *Christian Advocate*, 21 April 1887.
50. *Belfast Newsletter*, 26 March 1912.
51. Patterson, 'Independent Orangeism', p. 12.
52. On the Trades Council see Morgan, *Labour and Partition*, ch 4 and *passim*.
53. Belfast Trades Council Minute Book, 18 November 1905; 2 July 1908; 3 September 1908; 20 April 1912; 4 July 1912; 7 November 1912.
54. Johnstone, *The Vintage of Memory*, pp. 123ff; see also Larkin, *James Larkin*, pp. 26–38.
55. Greacen, *The Sash my Father Wore*, p. 24.
56. Green, *Religion in the Age of Decline*, p. 214.

57. See, for example, McCartney, *op. cit.*, p. 272.
58. Hanna Street Mission Sunday School Minute Book, 1st Annual Report: PRONI, D/2021/1.
59. See *Mission Completed*, p. 11.
60. Moody, *Memories and Musings*, p. 26.
61. *Belfast Telegraph*, 29 March 1969.
62. Kerr, *Rev Andrew Boyd*, p. 20.
63. The account in the *Dublin Evening Mail* (16 July 1887) was reprinted in ILPU leaflet no 41 (second series).
64. Barron, *The God of my Life*, pp. 212–18.
65. Cooney, *A History of the Christian Endeavour Movement in Ireland*, p. 11.
66. *Irish Endeavourer*, vol lxix, no 6, June 1963.
67. Cole, *History of Methodism in Ireland*, p. 81.
68. Springhall, *Sure and Stedfast*, p. 26.
69. Quoted in *ibid*, p. 25.
70. Kelly, *Firm and Deep*, p. 20.
71. Minute Book of the Church Lads' Brigade, Belfast Battalion, 30 September 1908: PRONI D/950/1/148.
72. Springall, *op. cit.*, p. 39.
73. See, for example, Bebbington, 'The City, the Countryside and the Social Gospel', in Baker, *The Church in Town and Countryside.*
74. Quoted in Springhall, *Coming of Age*, p. 139.
75. Quoted in Jeffery, *'An Irish Empire'*, p. 149.
76. *The Warden*, 19 August 1910.
77. Corkey, *David Corkey*, p. 91.
78. *Ibid*, p. 278.
79. See *The Growth and Development of the Girls' Brigade in Ireland*, p. 4.
80. See Holmes and Knox, *The General Assembly of the Presbyterian Church*, p. 120.
81. *Irish Churchman*, 19 April 1912.
82. Quoted in Gallagher, *At Points of Need*, pp. 25–6.
83. Parker, *For the Family's Sake*, p. 13.
84. Burrows and Mayes, *Specially Concerned*, p. 9.
85. Butler, *The Sub-Prefect Should Have Held his Tongue*, p. 1.
86. Selborne, *A Defence of the Church of England*, p. 81.
87. Gray, *The Orange Order*, pp. 193–4.
88. Quoted in Biggs-Davidson and Chowdharay-Best, *The Cross of St Patrick*, p. 250.
89. Holmes, *Presbyterians and Orangeism*, p. 10.
90. McCaughey, *Memory and Redemption*, p. 28.
91. Jarman, 'Displaying Faith', p. 12.
92. McClelland, 'Orange Folk Art in Ireland', p. 15.
93. Bowen, *Protestants in a Catholic State*, p. 187.
94. Hartford, *Godfrey Day*, p. 133.
95. *Irish Churchman*, 27 June 1929.
96. *Ibid*, 21 June 1928.
97. *Ibid*, 1 May 1930.
98. Hastings, *A History of English Christianity*, p. 644.
99. *East Antrim Times*, 29 April 1977.

Chapter 4: Seeing and believing, 1900–1965

1. Mahaffy, 'Will Home Rule be Rome Rule?', p. 159.
2. Bullock, *After Sixty Years*, p. 62.
3. Bence-Jones, *Twilight of the Ascendancy*, p. 57.
4. *Ibid*, quoting Mary Ponsonby.
5. Nicolson, *Helen's Tower*, p. 248.
6. Diary entry, 5 May 1935; in Mansergh, *Nationalism and Independence*, p. 123; John Betjman however thought St Anne's 'the best modern cathedral in the British Isles', *Belfast Telegraph*, 11 April 1975.
7. MS Autobiography of the Revd Matthew Wilson McCaul: PRONI D/1893/23.
8. Bence-Jones, *op. cit.*, p. 57.
9. Dublin Presbytery Minutes, 11 Oct 1898: PHS.
10. *Irish Presbyterian*, May 1901.
11. *The Warden*, 30 July 1909.
12. Church of Ireland Conference (1910) Minute Book, p. 92: RCB 236.
13. See, for example, Brown, 'The Presbyterian Dilemma', p. 61.
14. Munson, *The Nonconformists*, p. 150.
15. *Ibid*, p. 137.
16. Gillespie and Kennedy, *Ireland: Art into History*, p. 214.
17. *Ibid*, p. 215.
18. R Cobain, 'The Burning Bush', in *Presbyterian Herald*, July/August 1987, p. 10.
19. *The Warden*, 6 November 1908.
20. See Galloway, *The Cathedrals of Ireland*.
21. See Bowe, *The Life and Work of Harry Clarke*.
22. Larmour, *The Arts and Crafts Movement in Ireland*, p. 168.
23. Gillespie and Kennedy, *op. cit.*, p. 212.
24. Machin, 'The Last Victorian Anti-Ritualist Campaign 1895–1906', p. 301.
25. O'Connor, *The Image of a Cross*.
26. *Ibid*, pp. 39–40.
27. Dudley Edwards, *The Faithful Tribe*, p. 224.
28. Irish Church Union, Annual Reports for 1935 and 1936: RCB 450.
29. Wilkinson, *The Church of England and the First World War*, p. 190.
30. *Public Opinion*, 29 October 1915.
31. Bernard, *In War Time*, p. 76.
32. *Witness*, 17 September 1915.
33. Hartford, *Godfrey Day*, p. 45.
34. *JGS, 1930*, p. lxxiv.
35. Hylson-Smith, *High Churchmanship*, p. 273.
36. Seaver, *John Allen Fitzgerald Gregg*, p. 157; for these two cases see ch IX.
37. See Irish Church Union Papers, 1933–50: RCB 450.
38. Warke, *Ripples in the Pool*, p. 45.
39. *JGS, 1964*, p. cxiv.
40. *Belfast Telegraph*, 7 September 1964.
41. *Ulster Protestant*, February 1969.
42. Rattenbury, *The Eucharistic Hymns of John and Charles Wesley*, 22.
43. Manning, *The Hymns of Wesley and Watts*, p. 83.

44. Kenny, *Goodbye to Catholic Ireland*, p. 102.
45. Bradshaw, *Blood Worship: Christian Idolatry* (probably published 1908–14).
46. Bradley, *Abide with Me*, p. xvi.
47. *Methodist Newsletter*, July/August 1992.
48. Quoted in Jeffery, 'Four Methodist Hymn Books', p. 19.
49. See Davey, *The Story of a Hundred Years*, pp. 57–8.
50. Barkley, 'Marriage and the Presbyterian Tradition', p. 33.
51. Corkey, *David Corkey*, p. 37.
52. Blair, 'The Rev W.F. Marshall', p. 2.
53. *Presbyterian Herald*, September 1971.
54. Holmes and Knox, *The General Assembly*, p. 144.
55. Maguire, *Fifty-Eight Years*, p. 39.
56. Murray, *How can we Make our Churches and Services More Attractive*, p. 16.
57. Collis, *The Sparrow hath Found Herself a House*, p. 83.
58. *Irish Churchman*, 29 November 1928.
59. *Heritage and Renewal: The Report of the Archbishops' Commission on Cathedrals*, pp. 187 ('Note on the Historical Background' by Edward Norman).
60. Grindle, *Irish Cathedral Music*, pp. 125–6.
61. *Ibid*, p. 123.
62. Richardson, *A Tapestry of Beliefs*, p. 77.
63. Whiteside, *George Otto Simms*, p. 9.
64. Bullock, *op. cit.*, p. 69.
65. Diary of the Revd W. Moutray: PRONI D/2023/7/1/45 and 47.
66. Moore, *Reminiscences and Reflections*, p. 263.
67. Fitzpatrick, *Oceans of Consolation*, pp. 462–3.
68. Davies, *Worship and Theology in England*, vol 4, p. 284.
69. Allen, *The Presbyterian College, Belfast*, p. 184.
70. Murray, *op. cit.*, p. 13.
71. Grindle, *op. cit.*, p. 96.
72. *Irish Churchman*, 2 March 1933.
73. Leslie, *Long Shadows*, p. 44.
74. Patton, *Fifty Years of Disestablishment*, pp. 248–9; C of I Conference Minute Book, p. 92: RCB 236.
75. Harford and MacDonald, *Handley Carr Glyn Moule*, p. 254.
76. Hylson-Smith, *op. cit.*, p. 280.
77. Butler, 'Select Documents XLV: Lord Oranmore's Journal', p. 582.
78. Iremonger, *William Temple*, p. 378.
79. Graham, *Just As I Am*, p. 429.
80. Johnstone, *Vintage of Memory*, p. 157.
81. Barnes, *All for Jesus*, p. 71.
82. *Ibid*, p. 104.
83. Douglas, *Twentieth Century Dictionary of Christian Biography*, p. 275.
84. Redmond, *Church, State, Industry*, p. 64.
85. McCartney, *Nor Principalities Nor Powers*, p. 292.
86. Sibbett, *For Christ and Crown*, p. 322.
87. *Witness*, 25 February 1927.
88. *Irish Churchman*, 19 April 1928.
89. Cooke, *Persecuting Zeal*, p. 36.

90. On the church and its pastor see McCreedy, *The Seer's House*.
91. Graham, *op. cit.*, pp. 427–30.
92. Smiley, *Life and Letters of the Revd W. Smiley*, p. 79.
93. Bleakley, *Young Ulster and Religion*, p. 47.
94. The Very Revd Robert McCarthy, *Church Times*, 17 September 1999.
95. *Belfast Telegraph*, 4 March 1983.
96. The Revd W.L. Northridge; interview with the author, 1965.
97. Johnstone, *op. cit.*, p. 118.
98. Harte, *The Road I Have Travelled*, p. 36.
99. Stanford and McDowell, *Mahaffy*, p. 127.
100. *Presbyterian Herald*, May 1984.
101. *Belfast Telegraph*, 14 August 1969.
102. Cole, *History of Methodism in Ireland*, p. 46.
103. Sibbett, *op. cit.*, photograph opposite p. 224.
104. Hilliard, 'Unenglish and Unmanly', pp. 185 and 190.
105. See Kenny, *Goodbye to Catholic Ireland*, p. 45.
106. Sibbett, *op. cit.*, photograph opposite p. 288.
107. McCartney, *op. cit.*, photograph opposite p. 305.
108. See Galatians 5, 16–21.
109. *Christian Advocate*, 30 March 1894.
110. *MCM, 1905*, p. 86; 1962, p. 108.
111. Johnstone, *op. cit.*, pp. 142–3.
112. *Irish Presbyterian*, September 1900.
113. *Ibid*, July 1901.
114. Birmingham, *Pleasant Places*, p. 58.
115. *The Warden*, 12 August 1910.
116. *Ibid*, 6 November 1908.
117. Hartford, *Godfrey Day*, p. 88.
118. Hartford, *Among the Clergy*, p. 95.
119. *The Times*, 15 December 1998.
120. Birmingham, *op. cit.*, p. 12.
121. Hughes, *Hugh Price Hughes*, p. 340.
122. *Irish Presbyterian*, March 1901.
123. Chadwick, *The Victorian Church Part II*, p. 423.
124. Bebbington, *Evangelicalism in Modern Britain*, p. 209.
125. *Ibid*, p. 263.
126. O'Driscoll, *The Leap of the Deer*, p. 126.
127. Corkey, *David Corkey*, p. 102.
128. *Ibid*; photograph opposite p. 80.

Chapter 5: Confessional states, 1922–1965

1. *JGS, 1921*, p. lii.
2. *Ibid, 1922*, p. liii.
3. *GAR, 1921*, pp. 70–1.
4. Buckland, *A History of Northern Ireland*, p. 56.
5. Akenson, *Education and Enmity*, p. 54.
6. *Witness*, 12 August 1921.

7. Quoted in Farren, *The Politics of Irish Education*, p. 47.
8. *Witness*, 1 September 1922.
9. Cabinet Conclusions, 15 December 1922: PRONI CAB 4/61.
10. Lynn to Londonderry, 19 February 1923: PRONI Lynn Papers D/3480/59/44.
11. *Witness*, 23 March 1923.
12. *Ibid*, 20 April 1923.
13. Corkey, *Episode in the History of Protestant Ulster*, p. 22.
14. Cabinet Conclusions, 27 September 1923: PRONI CAB 4/86.
15. Corkey, *op. cit.*, p. 39.
16. Akenson, *op. cit.*, p. 83.
17. *Ibid*, p. 88.
18. Farren, *op. cit.*, p. 78.
19. Corkey, *op. cit.*, pp. 68–9.
20. Cabinet Conclusions, 26 March 1929: PRONI CAB 4/229.
21. *Irish Churchman*, 22 May 1930.
22. *GAM, 1930*, p. 48.
23. *Ibid, 1922*, pp. 48 and 52.
24. *Ibid, 1931*, pp. 58–9.
25. Corkey, *op. cit.*, p. 93.
26. *Ibid*, p. 98.
27. Craigavon to Corkey, 21 August 1931: PRONI CAB 4/290.
28. *Belfast Telegraph*, 18 December 1931.
29. *Belfast Newsletter*, 23 March 1932.
30. *GAR, 1932*, pp. 71–2.
31. Corkey, *op. cit.*, p. 101.
32. *JGS, 1940*, pp. lxxvi–lxxviii (7 May 1940).
33. *Hansard (N.I.)*, vol 23, col 1304 (29 May 1940).
34. *GAR, 1941*, p. 48.
35. Butler, *The Art of the Possible*, pp. 92–3.
36. See Corkey, *op. cit.*, p. 105; Waddell, *John Waddell*, pp. 182–3.
37. Corkey, *op. cit.*, pp. 105–6.
38. *GAM, 1944*, p. 51; Waddell, *op. cit.*, p. 184.
39. *Irish Christian Advocate*, 3 March, 10 March 1944.
40. Corkey, *op. cit.*, p. 118.
41. *JGS, 1946*, p. cxix.
42. Seaver, *John Allen Fitzgerald Gregg*, p. 248.
43. Waddell, *op. cit.*, p. 185.
44. *GAM, 1947*, p. 23.
45. *Irish Christian Advocate*, 1 and 22 November 1946; *MCM, 1946*, p. 74; *1947*, p. 72.
46. Akenson, *op. cit.*, p. 177.
47. See Birrell and Murie, *Policy and Government in Northern Ireland*, pp. 37 and 43.
48. Quoted in Akenson, *op. cit*, p. 177.
49. *GAR, 1921*, p. 71.
50. *Witness*, 23 February 1923.
51. *Ibid*, 18 May 1923.
52. Ervine, *Craigavon: Ulsterman*, p. 486.
53. Hyndman to Craig, 31 May 1923: PRONI CAB 8B/2.

54. *GAM, 1924*, p. 37.
55. *Witness*, 22 October 1926.
56. *Ibid*, 19 November 1926.
57. Letter dated 6 December 1924; Braley, *Letters of Herbert Hensley Henson*, p. 35.
58. See Longmate, *The Water-Drinkers*, p. 219.
59. *Irish Presbyterian*, December 1927.
60. *Irish Churchman*, 17 January 1929.
61. *Ibid*, 7 February 1929.
62. Galliher and DeGregory, *Violence in Northern Ireland*, p. 194.
63. Kenny, *Goodbye to Catholic Ireland*, p. 149.
64. *Irish Churchman*, 14 August 1930.
65. Legerton, *For our Lord and His Day*, p. 303.
66. Devlin, 'The Eucharistic Procession of 1908', p. 408.
67. Harris to Blackmore, 13 March 1931; Cabinet Secretariat File: 'International Eucharistic Congress, Dublin 1932': PRONI CAB 9B/200.
68. Fleming, *Head or Harp?*, p. 136.
69. Livingstone, *The Fermanagh Story*, p. 336.
70. *Irish Churchman*, 4 August 1932.
71. See Farmar, *Ordinary Lives*, p. 128.
72. Letter dated 1 July 1932; Cabinet Secretariat File: 'International Eucharistic Congress, Dublin 1932': CAB 9B/200.
73. Coogan, *De Valera*, p. 453.
74. Buckland, *op. cit.*, p. 79.
75. See Walker, 'Protestantism before Party!' for an examination of the UPL and the UPS.
76. Obituary in *MCM, 1957*, p. 12.
77. See Bowen, *History and the Shaping of Irish Protestantism*, p. 409.
78. *GAR, 1936*, p. 4.
79. Printed in *JGS, 1936*, pp. 370–80.
80. Sermon inaugurating the Diocesan Council of Youth, 3 May 1933; in MacNeice, *Some Northern Churchmen*, p. 55.
81. Walker, *op. cit.*, p. 961.
82. Norman, *Church and Society in England*, pp. 392–3.
83. Kenny, *op. cit.*, p. 201.
84. Harkness, *Ireland in the Twentieth Century*, pp. 82–5.
85. *Ibid*, p. 92.
86. Brodie, *The Tele*, p. 87.
87. Cabinet Conclusions, 11 March 1943: PRONI CAB 4/534.
88. *The Times*, 9 October 1999.
89. Minutes of Executive Committee, TCD Mission, 17 November 1939: RCB 478.
90. Cole, *History of Methodism in Ireland*, p. 141.
91. From an Order of Service leaflet of Holy Trinity Church, Portrush, 28 July 1940: RCB 515.
92. See discussion in Fisk, *In Time of War*, pp. 496–7.
93. *JGS, 1949*, p. lxxxiii.
94. Larkin, 'Church, State and Nation', p. 1276.
95. 21 September 1923, in Bowen, *Protestants in a Catholic State*, p. 34.

96. Foster, *Modern Ireland*, p. 449.
97. French, 'J.O. Hannay and the Gaelic League', p. 46.
98. Dublin Presbytery Minutes, 4 April 1905: PHS.
99. *Gaelic Churchman*, March 1919.
100. *Ibid*, November 1921.
101. Farren, *op. cit.*, p. 52.
102. Moore, *Reminiscences and Reflections*, pp. 287–8.
103. *JGS, 1923*, p. 216.
104. *Irish Times*, 17 November 1926: in Hurley, *Irish Anglicanism*, p. 144.
105. Bowen, *Protestants in a Catholic State*, p. 60.
106. *Irish Times*, 23 November 1929: in Jones, 'The Attitudes of the Church of Ireland', p. 76.
107. *Irish Churchman*, 11 July 1929.
108. Farren, *op. cit.*, p. 144.
109. Jones, *op. cit.*, p. 77.
110. Farren, *op. cit.*, p. 189.
111. *Irish Times*, 22 February 1913.
112. Quoted in Kenny, *op. cit.*, p. 147.
113. *Irish Churchman*, 24 April 1930.
114. *GAM, 1931*, p. 44.
115. *Irish Churchman*, 14 April 1932; what would they have made of the widely distributed 1999 film entitled *The Spy Who Shagged Me*?
116. *GAR, 1932*, p. 38.
117. *Ibid, 1945*, p. 74.
118. Gallagher, *At Points of Need*, p. 88.
119. *The Times*, 17 December 1977.
120. Adams, *Censorship: The Irish Experience*, pp. 24–6.
121. Harmon, *Sean O'Faolain*, p. 105.
122. Fitzpatrick, *The Two Irelands*, p. 230.
123. *Irish Churchman*, 1 September 1932.
124. Fitzpatrick, *op. cit.*
125. *Witness*, 8 June 1928.
126. *Irish Churchman*, 11 December 1930.
127. *Irish Times*, 25 February 1929; quoted in Hurley, *Irish Anglicanism*, p. 144.
128. *GAM, 1931*, p. 63.
129. See for example *MCM, 1939*, pp. 75–6 and 104.
130. Stanford, *A Recognised Church*, p. 16.
131. Kohn, *The Constitution of the Irish Free State*, p. 356.
132. *JGS, 1930*, p. 208; also *Irish Churchman*, 13 November 1930.
133. Coogan, *op. cit.*, pp. 458–9.
134. Lee, *Ireland*, pp. 204–5; Whyte, *Church and State*, p. 48.
135. See Faughnan, 'The Jesuits and the Drafting of the Irish Constitution'.
136. Seaver, *op. cit.*, p. 128.
137. Coogan, *op. cit.*, p. 153.
138. Obituary in *GAR, 1955*, p. 173.
139. Lee, *op. cit.*, p. 211.
140. Bowen, *op. cit.*, p. 35.
141. Stanford, *op. cit.*, p. 16.
142. Hurley, *Irish Anglicanism*, p. 146; also Kennedy, *The Widening Gulf*, p. 148.

143. Fleming, *op. cit.*, p. 170.
144. Blanchard, *The Irish and Catholic Power*, p. 68.
145. Feeney, *John Charles McQuaid*, p. 27.
146. Blanchard, *op. cit.*, p. 80.
147. 12 April 1951; see McKee, 'Church–State Relations', p. 191.
148. Lee, *op. cit.*, p. 321.
149. Inglis, *West Briton*, p. 165.
150. Hurley, *op. cit.*, p. 150.
151. Barkley, *Blackmouth and Dissenter*, p. 80.
152. *Ibid.*
153. Barkley, 'Marriage and the Presbyterian Tradition', p. 38.
154. Hurley, *op. cit*, p. 148; also Blanchard, *op. cit.*, pp. 171–2.
155. Inglis, *op. cit.*, p. 158.
156. Bowen, *History and the Shaping of Irish Protestantism*, p. 302.
157. Seaver, *op. cit.*, p. 312.
158. *JGS, 1951*, pp. 270–1.
159. Whiteside, *George Otto Simms*, p. 69.
160. Coogan, *op. cit.*, p. 669.
161. See Boycott, *Boycott.*
162. Butler, *The Sub-Prefect Should Have Held his Tongue*, pp. 95 and 98.
163. *Dáil Debates* (4 July 1957), quoted in Hurley, *op. cit.*, p. 151.
164. Lee, *op. cit.*, p. 340.
165. Warke, *Ripples in the Pool*, p. 121.
166. *Ibid*, p. 23.
167. *JGS, 1959*, p. lxxxix.
168. Bowen, *op. cit.*, p. 427.

Chapter 6: A century at home and abroad

1. Gairdner, *'Edinburgh 1910'*, pp. 91–2.
2. Norman, *The Catholic Church in Ireland*, p. 282.
3. *Christian Advocate*, 19 February 1886; again on 26 March it declared: 'They are in absolute unison, they think and speak alike.'
4. *GAM, 1888*, p. 533.
5. *Irish Ecclesiastical Gazette*, 19 May 1890.
6. *GAM*, 1895, p. XX.
7. See Benson, *The Life of Edward White Benson*, p. 522.
8. *Witness*, 24 January 1902.
9. *Ibid*, 20 June 1900; see also *ibid*, 12 January 1900.
10. *Ibid*, 9 November 1900.
11. Revd Dr William Park: in Holmes and Knox, *The General Assembly*, p. 204.
12. Macaulay, *Ireland in 1872*, p. 411.
13. *JGS, 1911*, p. l, and generally pp. l–lv.
14. *The Warden*, 11 June 1909.
15. Minute Book of the Church of Ireland Conference 1910, p. 23: RCB 236.
16. Crozier to Traill, 24 November 1913: PRONI Mic 87.
17. *Irish Churchman*, 23 December 1911.
18. The text is given in *JGS, 1915*, pp. 403–4.

19. Ellis, *Vision and Reality*, p. 13.
20. Bernard to Davidson, 30 April 1918: BL Add MS 52783.
21. Moneypenny, *The Two Irish Nations*, p. 17.
22. *GAM, 1917*, p. 394.
23. *Ibid, 1919*, p. 910.
24. *Ibid, 1921*, p. 21.
25. *Irish Churchman*, 20 September 1928.
26. *JGS, 1928*, p. liv.
27. *Ibid, 1929*, p. lvii.
28. *Irish Churchman*, 7 November 1929.
29. Henson to D'Arcy, 21 October 1931, in Braley, *Letters of Herbert Hensley Henson*, p. 63.
30. *GAM, 1931*, p. 42.
31. *JGS, 1931*, p. xx.
32. See Ellis, *op. cit.*, pp. 77–8.
33. *JGS, 1934*, p. lvix.
34. Seaver, *John Allen Fitzgerald Gregg*, p. 173.
35. Henson to D'Arcy, 21 October 1931, in Braley, *op. cit*, p. 63.
36. *Irish Churchman*, 11 January 1934.
37. *Ibid*, 5 October 1933.
38. *Irish Presbyterian*, July 1934.
39. Details of the debate (and Gregg's interventions) are in Seaver, *op. cit.*, pp. 176–9.
40. *GAM, 1935*, pp. 35–6.
41. Hurley, *Irish Anglicanism*, p. 75.
42. Seaver, *op. cit.*, p. 179.
43. See *GAM, 1937*, pp. 38–40; *MCM, 1938*, pp. 97–8.
44. Gregg, *Reunion*, p. 5.
45. *MCM, 1944*, p. 113.
46. *GAR, 1945*, p. 41.
47. Quoted in Holmes and Knox, *op. cit.*, p. 219.
48. *Lambeth Conference 1948*, p. 37.
49. *Ibid*, p. 40.
50. *The Lambeth Conference 1958*, section 2.20.
51. Patterson, *Over the Hill*, p. 11.
52. *Ibid*, p. 12.
53. Holmes and Knox, *op. cit.*, p. 228.
54. *Belfast Telegraph*, 11 May 1971.
55. Richardson, *A Tapestry of Beliefs*, p. 281.
56. *Methodist Newsletter*, June 1999.
57. Buchanan, 'In Retrospect: Alan Alexander Buchanan', pp. 38–9.
58. *Belfast Telegraph*, 21 April 1973.
59. Patterson, *op. cit.*, p. 38.
60. Dunlop, *A Precarious Belonging*, pp. 15–16.
61. *Ibid*, p. 42.
62. Gallagher and Worrall, *Christians in Ulster*, p. 134.
63. Holmes and Knox, *op. cit.*, p. 244; see also Gallagher, 'Northern Ireland – The Record of the Churches', p. 173.
64. Edited by the Revds John Ellison and G.H.S. Walpole.

65. Jeffery, *'An Irish Empire'?*, p. 18.
66. Details on http://www.leprosymission.org.
67. *Chota Nagpur Quarterly*, vol II, no 2 (May 1898), p. 36.
68. Mahto, *Hundred Years of Christian Missions*, p. 173.
69. *Ibid*, p. 174.
70. Morris, 'The Dublin University Mission', p. 313.
71. Maxwell, *Belfast's Halls of Fame*, p. 115.
72. See Carmichael, *Thou Givest ... They Gather.*
73. Houghton, *Amy Carmichael*, p. 249.
74. *Ibid*, p. 341.
75. Candidate Sub-committee Minutes, 16 March 1899: RCB 315/1.9/1.
76. *Ibid*, 19 May 1899.
77. *Ibid*, 13 June 1900.
78. Harford-Battersby, *Pilkington of Uganda*, p. 23.
79. *JGS, 1919*, pp. 236–41.
80. *Ibid*, p. 123.
81. By Robin Boyd in Thompson, *Into all the World*, pp. 33–4.
82. *GAR, 1937*, p. 16.
83. Boyd, *The Path of Valour*, p. 26.
84. Fulton, *Through Earthquake, Wind and Fire*, p. 51.
85. *Ibid*, ch 3, for a full examination of the process.
86. *Ibid*, p. 242.
87. *JGS, 1951*, pp. 243–5.
88. Thompson, *op. cit.*, p. 44.
89. *GAR, 1968*, p. 95.
90. Gibbs, *The Anglican Church in India*, p. 409.
91. Patterson, *op. cit.*, p. 16.
92. Thompson, *op. cit.*, p. 47.
93. *Belfast Telegraph*, 10 March 1973.
94. World Council of Churches, *World Conference on Church and Society: Official Report.*
95. Quoted in Paton, *Breaking Barriers*, p. 232.
96. *Ibid*, pp. 54–5.
97. Norman, *Christianity and the World Order*, p. 2.
98. Standing Committee Report 1999, http://www.ireland.anglican.org.
99. Tolhurst, *A Companion to 'Veritas Splendor'*, ch 2 and p. 63.
100. *Presbyterian Herald*, July/August 1999.
101. Computed from Leslie, *Clergy of Connor.*
102. Taggart, *The Irish in World Methodism*, p. 8.
103. *Ibid*, p. 17.
104. Thompson, *op. cit.*, p. 206.
105. Douglas, 'A Light to the Nations', p. 481.
106. *Church Times*, 24 July 1998; 14 August 1998.
107. Akenson, *Small Differences*, p. 45.
108. Kitson Clark, *An Expanding Society*, p. 3.
109. *Ibid*, p. 5.
110. Maguire, 'A Socio-Economic Analysis of Dublin', p. 54.
111. *MCM, 1886* (Pastoral Address).
112. *Christian Advocate*, 2 April 1886.

113. Miller, *Emigrants and Exiles*, p. 378.
114. *GAM, 1895*, p. 993.
115. Milne, *Protestant Aid*, p. 19.
116. Dublin Prison Gate Mission Minute Book, 14 September 1906: RCB 263/2.
117. *GAR, 1924*, p. 21.
118. *Irish Presbyterian*, September 1926.
119. *Irish Churchman*, 31 July 1930.
120. Thompson, *op. cit.*, p. 169.
121. *GAR, 1934*, p. 33.
122. Calculations based on: Rankin, *Clergy of Down and Dromore*; Leslie, *op. cit.*; Thompson, 'Irish Baptist College List of Students', 1892–1963, and 1964–1991.
123. Knox, 'The Irish Contribution to English Presbyterianism', p. 23.
124. Canavan, 'The Future of Protestantism', p. 240.
125. McDermott and Webb, *Irish Protestantism Today and Tomorrow*.
126. Compton, 'An Evaluation of the Changing Religious Composition', p. 204.
127. *JGS, 1939*, p. lxxviii.
128. McKinney and Sterling, *Gurteen College*, p. 1.
129. *Irish Times*, 10 June 1939.
130. *Ibid*, 8 August 1972.
131. Dunlop, *op. cit.*, p. 34.
132. *Irish Times*, 9 February 1983.
133. Considère-Charon, 'The Church of Ireland', p. 114.
134. *Belfast Telegraph*, 22 November 1971.
135. See Flanagan, 'The Shaping of Irish Anglican Secondary Schools'.
136. See Hime, *Home Education*.
137. White, *The Anglo-Irish*, p. 264.
138. Ervine, *Craigavon: Ulsterman*, pp. 44–5.
139. Cathcart, *The Most Contrary Region*, p. 46.
140. *Protestant Telegraph*, 22 February 1969.
141. Mahaffy, 'Provincial Patriotism', p. 1034.
142. *Belfast Newsletter*, 26 July 1973.
143. *Irish Times*, 25 July 1998.
144. *MCR, 1992*, p. 127.
145. Quoted in Bew *et al.*, *Passion and Prejudice*, p. 4.
146. O'Connor, *In Search of a State*, p. 142.
147. Bell, *Back to the Future*, p. 328.
148. Elliott, 'Religion and Identity in Northern Ireland', p. 153.
149. Compton, 'Demography – The 1980s in Perspective', p. 167; see also his 'An Evaluation of the Changing Religious Composition of the Population of Northern Ireland'.
150. Anderson and Shuttleworth, 'Sectarian readings of sectarianism', p. 93.
151. Dunlop, *op. cit.*, p. 82.

Chapter 7: Changing times, 1960–1975

1. Carson, *Riots and Religion*, p. 1.
2. Keane, *Letter to Daniel*, p. 222.
3. Kenny, *Goodbye to Catholic Ireland*, p. 283.

4. *Irish Times*, 2 March 1963, in Farmar, *Ordinary Lives*, p. 170.
5. *Linen Hall Review*, Spring 1988.
6. Farmar, *op. cit.*, p. 153.
7. Feeney, *John Charles McQuaid*, p. 45.
8. *Ibid*, p. 60.
9. *Irish Times*, 29 June 1962.
10. *Ibid*, 8 March 1966.
11. *Ibid*, 22 February 1966.
12. Quoted in O'Neill, *The Autobiography of Terence O'Neill*, p. 67.
13. *JGS, 1961*, p. xc.
14. *GAM, 1961*, p. 36.
15. *MCM, 1981*, p. 132.
16. Brown, *The Word of God among all Nations*, p. 118.
17. Paisley, *Are We to Lose our Protestant Heritage For Ever?*, p. 5.
18. See Cooke, *Persecuting Zeal*, p. 127.
19. *GAM, 1927*, p. 43.
20. Allen, *The Presbyterian College, Belfast*, p. 256.
21. Fulton, *J. Ernest Davey*, pp. 32–3.
22. Barkley, *Blackmouth and Dissenter*, p. 34.
23. Richardson, *A Tapestry of Beliefs*, p. 190; in ch 17 by Revd Dr R.C. Beckett.
24. Cooke, *op. cit.*, chapter 6.
25. The whole story is recounted in Moloney and Pollak, *Paisley*, pp. 67–74; see also T.E. Utley in *Sunday Telegraph*, 14 October 1973.
26. *Whither Methodism?*, April 1965.
27. *Ibid*, April–June 1968.
28. *Ibid*, January–March 1969.
29. *Ibid*, January–March 1970.
30. *Ibid*, April–June 1970.
31. See Cole, *History of Methodism in Ireland*, p. 100; Livingstone, *The Fermanagh Story*, p. 234.
32. Richardson, *op. cit.*, pp. 142ff.
33. Issues of *Ulster Christian* often include features on such churches or ventures.
34. Gallagher and Worrall, *Christians in Ulster*, p. 38.
35. *Whither Methodism?*, April–June 1968.
36. *Irish Presbyterian*, March 1905.
37. Dillon, *God and the Gun*, p. 94.
38. *Belfast Telegraph*, 3 October 1964.
39. *Ibid*, 2 July 1970.
40. *JGS, 1920*, p. lxiii.
41. *Irish Churchman*, 5 September 1929.
42. *JGS, 1920*, p. lxix.
43. *Irish Churchman*, 6 November 1930.
44. See Holmes and Knox, *The General Assembly*, p. 124.
45. *GAM, 1930*, p. 26.
46. *JGS, 1949*, p. cxxvii.
47. *Ibid*, 21 June 1972.
48. *GAM, 1973*, p. 28.
49. *MCM, 1974*, p. 41.

50. *Ibid, 1975*, p. 39.
51. Quinn et al., *Directions*, pp. 178–9.
52. Acheson, *A History of the Church of Ireland*, p. 245.
53. *The Times*, 6 December 1996.
54. O'Dowd and Wichert, *Chattel, Servant or Citizen*, pp. xi–xii.
55. Kenny, *op. cit.*, p. 293.
56. Richardson, *op. cit.*, p. 227.
57. *Belfast Telegraph*, 29, 30, 31 January 1964; Patterson, *Over the Hill*, p. 31.
58. *Belfast Newsletter*, 3 October 1964.
59. *Belfast Telegraph*, 7 October 1964.
60. Bew *et al.*, *Passion and Prejudice*, p. 110 and *passim*.
61. Barritt and Carter, *The Northern Ireland Problem*, p. 154.
62. *GAR, 1964*, pp. 114–22.
63. *Ibid, 1965*, pp. 116–29.
64. *GAM, 1965*, p. 26.
65. *Ibid, 1967*, p. 59.
66. *GAR, 1967*, pp. 120–30.
67. *Presbyterian Herald*, July/August 1966.
68. *MCM, 1966*, pp. 135–7.
69. *Belfast Telegraph*, 1 March 1966.
70. *Ibid*, 12 July 1966.
71. *Ibid*, 11 July 1966.
72. Gallagher and Worrall, *op. cit.*, p. 36.
73. Moloney and Pollak, *Paisley*, p. 146.
74. *Ibid*; the guest is not named.
75. Gallagher and Worrall, *op. cit.*, p. 37.
76. Moloney and Pollak, *op. cit.*, p. 150.
77. *Protestant Telegraph*, 14 December 1968.
78. Utley, *Lessons of Ulster*, p. 41.
79. The ICC Report is printed in *GAR, 1969*, pp. 63–4.
80. Gallagher and Worall, *op. cit.*, p. 42.
81. *Belfast Telegraph*, 23 December 1968.
82. *Ibid*, 12 December 1968.
83. *Belfast Newsletter*, 14 January 1969.
84. *Belfast Telegraph*, 10 December 1968.
85. *Ibid*, 18 November 1968.
86. *GAR, 1969*, pp. 5–11.
87. *Belfast Telegraph*, 1 February 1969.
88. *Ibid*, 10 December 1968.
89. *JGS, 1969*, p. xlvii.
90. *GAM, 1969*, p. 26.
91. *MCM, 1969*, p. 88.
92. *Belfast Telegraph*, 7 August 1969.
93. *Irish Times*, 15 August 1969.
94. *Belfast Telegraph*, 5 August 1969.
95. *Ibid*, 16 August 1969.
96. *Ibid*, 18 September 1969.
97. *Belfast Newsletter*, 18 August 1969.
98. Whiteside, *George Otto Simms*, p. 119.

99. *Belfast Telegraph*, 18 August 1969.
100. *Ibid*, 20 August 1969.
101. Faulkner, *Memoirs of a Statesman*, p. 63.
102. Of the 14 Irish bishoprics, all but Kilmore, Meath, Cork and Ossory changed hands in 1969–72.
103. MacStiofáin, *Revolutionary in Ireland*, p. 156.
104. *Belfast Telegraph* and *Belfast Newsletter*, 1 September 1969.
105. ICC's report for 1969–70 printed in *JGS, 1970*, pp. 147–57.
106. *Belfast Telegraph*, 10 August 1971.
107. Whiteside, *op. cit.*, p. 131.
108. Flackes, *Northern Ireland: A Political Directory*, pp. 320–2.
109. Gallagher and Worrall, *op. cit.*, pp. 1–2, 69–70, 96–102.
110. Whitelaw, *The Whitelaw Memoirs*, p. 100.
111. MacStiofáin, *op. cit.*, p. 289.

Chapter 8: Towards the new millennium

1. Agrawal, *A Christian Response to the Irish Situation*, p. 42.
2. *Belfast Telegraph*, 11 April 1974.
3. *Ibid*, 7 January 1980.
4. *Observer*, 29 September 1991.
5. *The Times*, 15 November 1991.
6. *Sunday Times*, 18 March 1973.
7. *The Times*, 24 March 1988.
8. *Ibid*, 7 November 1990.
9. *Violence in Ireland*, p. 85.
10. *The Report of the Lambeth Conference 1978*, p. 51.
11. Whiteside, *George Otto Simms*, p. 148.
12. *The Times*, 2 June 1972.
13. His diocesan magazine quoted in *Belfast Telegraph*, 10 October 1972.
14. Quoted in *Irish Times*, 20 October 1972.
15. *Belfast Telegraph*, 2 May 1972.
16. Holden, 'Will the Inward Light go out in Ireland', p. 323.
17. See Bailey, *Acts of Union*, pp. 107–15.
18. *Belfast Telegraph*, 20 September 1972; *Sunday Times*, 30 September 1973.
19. *Belfast Newsletter*, 16 November 1973.
20. *Belfast Telegraph*, 9 July 1980.
21. Stringer and Robinson, *Social Attitudes in Northern Ireland*, p. 84.
22. Gilbert, *The Making of Post-Christian Britain*, p. 145.
23. *GAR, 1971*, p. 152 (Committee on National and International Problems).
24. *Ibid*, p. 171.
25. *Ibid*, p. 145.
26. *Irish Times*, 16 May 1970.
27. *JGS, 1971*, p. xlv.
28. *Belfast Telegraph*, 17 October 1973.
29. Norris, 'Homosexual People and the Christian Churches', p. 35.
30. *MCM, 1977*, p. 64.
31. Norris, *op. cit.*

32. Acheson, *A History of the Church of Ireland*, p. 245.
33. *Daily Telegraph*, 17 November 1994; *Belfast Telegraph*, 13 June 1995.
34. *Presbyterian Herald*, May 1977.
35. *Belfast Telegraph*, 6 July 1977.
36. *Marriage and the Family in Ireland To-day*, p. 19.
37. Kenny, *Goodbye to Catholic Ireland*, p. xxv.
38. *Violence in Ireland*, p. 86.
39. *Belfast Telegraph*, 25 April 1969.
40. *Ibid*, 17 April 1971.
41. *Ibid*, 6 December 1972.
42. Chadwick, *Michael Ramsey*, p. 204.
43. McIvor, *Hope Deferred*, p. 112.
44. *Ibid*, p. 113.
45. Faulkner, *Memoirs of a Statesman*, p. 242.
46. *Belfast Telegraph*, 1 May 1974.
47. Faulkner, *op. cit.*
48. *Belfast Telegraph*, 3 September 1971.
49. *Ibid.*
50. *Ibid*, 8 April 1974.
51. McCann, 'Christian Education in the Service of Reconciliation', p. 69.
52. *Belfast Telegraph*, 7 May 1990.
53. *GAM, 1974*, p. 52.
54. *JGS, 1974*, pp. 129–35.
55. Gallagher and Worrall, *Christians in Ulster*, p. 167.
56. Collins, 'Political Attitudes and Integrated Education', pp. 410–12.
57. Brown, 'Time to Rethink?', p. 53.
58. See Elliott, 'Religion and Identity in Northern Ireland', p. 157.
59. *Belfast Telegraph*, 27 September 1990.
60. Morgan and Fraser, 'Integrated Schools, Religious Education and the Churches', p. 34.
61. *Belfast Telegraph*, 17 June 1995.
62. Brown, *op. cit.*, p. 56.
63. *Irish Times*, 15 June 1982.
64. *Administration 1967*, p. 7.
65. *Ibid*, p. 123.
66. Paul, *The Deployment and Payment of the Clergy*.
67. Hastings, *A History of English Christianity*, pp. 535–6.
68. *Administration 1967*, p. 38.
69. See *Belfast Telegraph*, 24 June 1978.
70. Patterson, *Over the Hill*, p. 46; *GAM, 1974*, pp. 48 and 56.
71. Patterson, *Over the Hill*, pp. 38–51.
72. *GAM, 1982*, p. 56.
73. Thompson, 'The Churches in Northern Ireland', p. 269.
74. Warke, *Ripples in the Pool*, p. 121.
75. FitzGerald, *Father Tom*, p. 62.
76. *Belfast Telegraph*, 1 August 1978.
77. *Belfast Newsletter*, 7 August 1978.
78. *Ibid*, 8 August 1978.
79. *Belfast Telegraph*, 3 April 1986.

80. Eames, *Chains to be Broken*, pp. 52–3.
81. *Belfast Telegraph*, 18 April, 14 June 1986.
82. Bauman, *Mortality, Immortality*, pp. 188–9.
83. Warke, 'Baptists in Belfast', p. 19.
84. McCann, *op. cit.*, p. 66.
85. *Belfast Telegraph*, 28 March 1995.
86. *Ibid*, 14 May 1991.
87. *Ibid*, 30 September 1994.
88. *Frontiers*, 3:3, Spring 1999, p. 6.
89. *Church Times*, 27 January 1995.
90. See Stevens, 'Dealing with the Lethal Toxin of Sectarianism'.
91. Pastoral Address of Conference in *Methodist Newsletter*, July/August 1992.
92. *Church Times*, 30 January 1998.
93. See Hartin, 'The Challenge to Irish Ecumenism'.
94. Lapsley, 'Where is the Ecumenical Movement Going Today?', p. 300.
95. *Irish Times*, 19 November 1994.
96. *The Times*, 23 January 1995.
97. *Presbyterian Herald*, July/August 1999.
98. *Irish Times*, 30 June 1990.
99. McVeigh, 'The Specificity of Irish Racism', pp. 41–2.
100. Keane, *Letter to Daniel*, p. 147.
101. Murphy, *A Place Apart*, p. 281.
102. *Belfast Newsletter*, 1 May 1972.
103. McIvor, *Hope Deferred*, p. 96.
104. See *The Times*, 15 November 1976 on the 'possible creation of an independent republic of Ulster'; and 20 October 1978 on the UDA's plans for 'the creation of an independent state of Northern Ireland'.
105. *The Times*, 3 July 1981.
106. *Orange Standard*, December 1979/January 1980.
107. *Church Times*, 29 May 1998.
108. In Hughes, *Culture and Politics in Northern Ireland*, p. 22.
109. Campbell, *The Dissenting Voice*, p. 335.
110. See, for example, Belfrage, *Living with War*, ch 5; Murphy, *op. cit.*, pp. 58 and 222; McCrum, *The Craic*, ch 25.
111. Wheatcroft, 'Ireland and the Left', p. 11.
112. Revd Dr William Park in 1890; see above, pp. 122–3.

Appendix: Succession Lists

CHURCH OF IRELAND
ARCHBISHOPS & BISHOPS

Province of Armagh
Armagh [Primates and Archbishops]

1896 W Alexander (ex Derry)
1911 JB Crozier (ex Ossory; Down)
1920 CF D'Arcy (ex Clogher; Ossory, Down, Dublin)
1938 JGF Day (ex Ossory)
1939 JAF Gregg (ex Ossory; Dublin)
1959 J McCann (ex Meath)
1969 GO Simms (ex Cork, Dublin)
1980 JW Armstrong (ex Cashel)
1986 RHA Eames (ex Derry; Down)

Clogher *[united to Armagh 1850–86]*

1886 CM Stack
1903 CF D'Arcy (to Ossory)
1908 M Day
1923 J MacManway
1944 R Tyner
1958 AA Buchanan (to Dublin)
1970 RPC Hanson
1973 RW Heavener
1980 G McMullan
1986 BDA Hannon

Connor *[united to Down 1441–1945]*

1945 CK Irwin (ex Down)
1956 RCHG Elliott
1969 AH Butler (ex Tuam)
1981 WJ McCappin
1987 SG Poyntz (ex Cork)
1995 JE Moore
JE Moore

Derry [and Raphoe]

1896 GA Chadwick
1916 JI Peacocke
1945 RMcN Boyd (ex Killaloe)
1958 CJ Tyndall (ex Kilmore)
1970 CI Peacocke
1975 RHA Eames (to Down)
1980 J Mehaffy

Down [Connor and Dromore]

1892 TJ Welland
1907 JB Crozier (ex Ossory; to Armagh)
1911 CF D'Arcy (ex Clogher, Ossory; to Dublin)
1919 CTP Grierson
1934 JF McNeice (ex Cashel)
1942 CK Irwin (to Connor) *[1945 Down and Dromore]*
1945 WS Kerr
1955 FJ Mitchell (ex Kilmore)
1970 GA Quin
1980 RHA Eames (ex Derry; to Armagh)
1986 G McMullan (ex Clogher)
1996 HC Miller

Kilmore [Elphin and Ardagh]

1897 AG Elliott
1915 WR Moore
1930 AW Barton (to Dublin)
1939 AE Hughes
1950 FJ Mitchell (to Down)
1956 CJ Tyndall (to Derry)
1959 EFB Moore
1981 WG Wilson
1993 MHG Mayes

Tuam [Killala and Achonry]

1890 J O'Sullivan
1913 BJ Plunket (to Meath)

1920 AE Ross
1923 J Ort
1928 JM Harden
1932 WH Holmes (to Meath)
1939 JW Crozier
1958 AH Butler (to Connor)
1970 JC Duggan
1986 JRW Neill (to Cashel)
1998 R Henderson

Province of Dublin
Dublin [and Glendalough]
[Archbishops]

1897 JF Peacocke (ex Meath)
1915 JH Bernard (ex Ossory)
1919 CF D'Arcy (ex Clogher, Ossory; Down, to Armagh)
1920 JAF Gregg (ex Ossory; to Armagh)
1939 AW Barton (ex Kilmore)
1956 GO Simms (ex Cork; to Armagh)
1969 AA Buchanan (ex Clogher)
1977 HR McAdoo (ex Ossory)
1985 DAR Caird (ex Limerick, Meath)
1986 WNF Empey (ex Limerick, Meath)

Cashel

1900 HS O'Hara
1919 R Miller
1931 JF McNeice (to Down)
1935 TA Harvey
1958 WC De Pauley
1968 JW Armstrong (to Armagh)
[1977 Cashel and Ossory]
1980 NV Willoughby
1997 JRW Neill (ex Tuam)

Cork [Cloyne and Ross]

1894 WE Meade
1912 CB Dowse (ex Killaloe)
1933 WE Flewett
1938 RT Hearn
1952 GO Simms (to Dublin)
1957 RG Perdue (ex Killaloe)
1978 SG Poyntz (to Connor)
1988 A Warke
1999 WP Colton

Killaloe

1897 M Archdall
1912 CB Dowse (to Cork)
1913 TS Berry
1924 HE Patton
1943 RMcN Boyd (to Derry)
1945 H Webster
1953 RG Perdue (to Cork)
1957 HA Stanistreet
1972 E Owen
[1976 United to Limerick]

Limerick

1899 T Banbury
1907 R d'A Orpen
1921 HV White
1934 CK Irwin
1942 EC Hodges
1961 RW Jackson
1970 DAR Caird (to Meath)
[1976 Limerick and Killaloe]
1976 E Owen
1981 WHF Empey (to Meath)
1985 EF Darling

Meath[1]

1897 JB Keene
1919 BJ Plunket (ex Tuam)
1926 TGG Collins
1927 J Orr
1938 WH Holmes (ex Tuam)
1945 J McCann (to Armagh)
1959 RB Pike
[1976 Meath and Kildare]
1976 DAR Caird (ex Limerick, to Dublin)
1985 WNF Empey (ex Limerick, to Dublin)
1996 RL Clarke

1 The Diocese of Meath and Kildare was transferred to the Province of Dublin in 1976

Ossory [and Ferns]

1897 JB Crozier (to Down)
1907 CF D'Arcy (ex Clogher, to Down)
1911 JH Bernard (to Dublin)
1915 JAF Gregg (to Dublin)
1920 JGF Day (to Armagh)
1938 F Tichbourne
1940 JP Phair
1962 HR McAdoo (to Dublin)
[1977 united to Cashel]

PRESBYTERIAN CHURCH MODERATORS
[* = died in office]

1900 JH Hamilton (Dublin)
1901 J Heron (Belfast)
1902 JE Henry (Londonderry)
1903 K MacDermott (Belfast)
1904 S Prenter (Dublin)
1905 W McMordie (Kilkeel)
1906 W McKean (Belfast)
1907 J Davidson (Glasslough)
1908 I McIlveen (Belfast)
1909 JC Clarke (Galway)

1910 JH Murphy (Cork)
1911 J Macmillan (Belfast)
1912 H Montgomery (Belfast)
1913 WJ Macaulay (Portadown)
1914 J Bingham (Dundonald)
1915 TMcA Hamill (Belfast)
1916 T West (Antrim)
1917 J Irwin (Belfast)
1918 J McGranahan (Londonderry)
1919 JM Simms (Newtownards)

1920 H Patterson Glenn (Bray)
1921 WJ Lowe (Belfast)
1922 WG Strahan (Newry)
1923 G Thompson (Belfast)
1924 RW Hamilton (Lisburn)
1925 T Haslett (Ballymena)
1926 RK Hanna (Dublin)
1927 J Thompson (Derry)
1928 TA Smyth (Belfast)
1929 JL Morrow (Dublin)

1930 E Clarke (Strabane)
1931 JG Paton (Belfast)
1932 JJ Macaulay (Dublin)
1933 W Corkey (Belfast)
1934 TMcG Johnstone (Belfast)
1935 AF Moody (Belfast)
1936 WJ Currie (Bangor)
1937 J Waddell (Belfast)
1938 WJ Currie (Bangor)
1939 J Haire (Belfast)

1940 JB Woodburn (Belfast)
1941 WA Watson (Belfast)
1942 WM Kennedy (Derry)
1943 P McKee (Newry)*
1944 A Gibson (Cork)
1945 R Corkey (Belfast)
1946 T Byers (Dublin)
1947 R Boyd (Belfast)
1948 A W Neill (Armagh)*
1949 GD Erskine (Belfast)

1950 GD Erskine (Belfast)
1951 H McIlroy (Ryans, Newry)
1952 JKL McKean (Comber)
1953 JE Davey (Belfast)
1954 J Knowles (Tullylish)
1955 JC Breakey (Belfast)
1956 TM Barker (Donegal)
1957 RJ Wilson (Belfast)
1958 W McAdam (Newry)
1959 TAB Smyth (Dublin)

1960 AA Fulton (Belfast)
1961 WAA Park (Ballygilbert)
1962 JH Davey (India)
1963 WA Montgomery (Derry)
1964 J Dunlop (Belfast)
1965 SJ Park (Dublin)
1966 T McC Barker (Belfast)
1967 W Boyd (Lisburn)
1968 JH Withers (Belfast)
1969 JT Carson (Bangor)

1970 JLM Haire (Belfast)
1971 FR Gibson (Belfast)
1972 RVA Lynas (Larne)
1973 JW Orr (Belfast)
1974 GT Lundie (Armagh)
1975 DGF Wynne (Derry)*
1976 AJ Weir (Belfast)

1977 TA Patterson (Portaferry)
1978 D Burke (Bangor)
1979 WM Craig (Portadown)

1980 RG Craig (Carrickfergus)
1981 J Girvan (Lurgan)
1982 EP Gardner (Ballymena)
1983 TJ Simpson (Newtownards)
1984 H Cromie (Lisburn)
1985 R Dickinson (Tobermore)
1986 J Thompson (Belfast)
1987 W Fleming (Belfast)
1988 AWG Brown (Ballycastle)
1989 JA Matthews (Lurgan)

1990 RFG Holmes (Belfast)
1991 R Sterritt (Newtownards)
1992 J Dunlop (Belfast)
1993 AR Rodgers (Dungannon)
1994 DJ McGaughey (Kilkeel)
1995 J Ross (Hollywood)
1996 DH Allen (Coleraine)
1997 S Hutchison (Clerk)
1998 J Dixon (Antrim)
1999 J Lockington (Larne)

Clerks to the General Assembly

1895–1931 WJ Lowe
1931–1953 WA Watson
1953–1962 JHR Gibson
1962–1963 AJ Gailey*
1963–1985 AJ Weir
1985–1990 TJ Simpson
1990– S Hutchinson

METHODIST CHURCH PRESIDENTS

1900 W Crawford (and 1907)
1901 J Park (and 1910)
1902 W Guard (and 1893, 1911)
1903 W Nicholas (and 1894)
1904 T Knox
1905 GR Wedgwood (and 1912)
1906 J Robertson (and 1897)
1907 W Crawford (and 1900)
1908 JD Lamont
1909 JWR Campbell

1910 J Park (and 1901)
1911 W Guard (and 1893, 1902)
1912 GR Wedgwood (and 1905)
1913 ST Boyd
1914 WR Budd
1915 JO Price
1916 P Martin
1917 W Maguire
1918 H McKeag
1919 J Kirkwood

1920 H Shire
1921 WH Smyth (and 1927)
1922 JM Alley
1923 JW Parkhill
1924 W Corrigan
1925 EB Cullen
1926 RM Ker
1927 WH Smyth (and 1921)
1928 RC Phillips
1929 JC Cameron

1930 W Moore
1931 FE Harte
1932 JA Duke
1933 RL Cole
1934 JA Walton
1935 TJ Irwin
1936 WH Massey
1937 CH Crookshank
1938 TJ Allen
1939 A McCrea

1940 HM Watson
1941 JN Spence
1942 BS Lyons
1943 GA Joynt
1944 WL Northridge
1945 E Whittaker
1946 RH Gallagher
1947 J England
1948 WEM Thompson
1949 JW Stutt

1950 JRW Roddie
1951 HN Medd
1952 J Montgomery
1953 RML Waugh
1954 E Shaw
1955 A Holland

1956 SE McCaffrey
1957 JW McKinney
1958 RJ Good
1959 RE Ker
1960 RW McVeigh
1961 CW Ranson
1962 J Wisheart
1963 FE Hill
1964 SH Baxter
1965 RA Nelson
1966 SJ Johnston
1967 RDE Gallagher
1968 GG Myles
1969 GE Good
1970 J Davison
1971 CH Bain
1972 ER Lindsay
1973 H Sloan
1974 RD Morris
1975 HW Plunkett
1976 R Greenwood
1977 RG Livingstone
1978 J Turner
1979 V Parkin
1980 WS Callaghan
1981 EW Gallagher
1982 CG Eyre
1983 CA Newell
1984 P Kingston (B)[2]
1985 H Skillen
1986 S Frame
1987 WI Hamilton
1988 TS Whittington
1989 GR Morrison
1990 WT Buchanan
1991 JW Good
1992 JDG Ritchie
1993 RH Taylor
1994 ETI Mawhinney
1995 CG Walpole
1996 K Best
1997 NW Taggart
1998 DJ Kerr
1999 KA Wilson

2 This designation (B) is used to distinguish him from another, much younger, Revd P Kingston

Select Bibliography

Manuscripts

Full citations for documents quoted are given in the Notes, but the following abbreviations should be noted:

BL	British Library
CUL	Cambridge University Library
NLI	National Library of Ireland
PHS	Presbyterian Historical Society, Belfast
PRO	Public Record Office, London
PRONI	Public Record Office of Northern Ireland
RCB	Representative Church Body Library, Dublin
TCD	Trinity College Dublin Library
WHS	Wesley Historical Society (Irish Branch) Archives, Belfast

Newspapers and periodicals

These are cited fully in the Notes.

Reference

A History of Congregations in the Presbyterian Church in Ireland 1610–1982 (Belfast 1982).

Barkley, J.M. *Fasti of the General Assembly of the Presbyterian Church in Ireland (Part I) 1840–1870* (Belfast 1986); *(Part II) 1871–1890* (Belfast 1987); *(Part III) 1891–1910* (Belfast 1986).

Crockford's Clerical Directory (London, published annually).

Douglas, J.D. (ed.) *Twentieth Century Dictionary of Christian Biography* (Carlisle 1995).

Flackes, W.D. *Northern Ireland: A Political Directory* (London 1983).

Irish Church Directory (Dublin, published annually; latterly *The Church of Ireland Directory*).

Leslie, J.B. *Clergy of Connor from Patrician Times to the Present Day* (Belfast 1993).

Rankin, F. (ed.) *Clergy of Down and Dromore* (Belfast 1996).

Books

Only books cited in the Notes are listed here.

Acheson, A. *A History of the Church of Ireland 1691–1996* (Dublin 1997).

Adams, M. *Censorship: The Irish Experience* (Dublin 1968).

Akenson, D.H. *Education and Enmity: The Control of Schooling in Northern Ireland 1920–1950* (Newton Abbot 1973).
Akenson, D.H. *Small Differences: Irish Catholics and Irish Protestants 1815–1922* (Montreal 1988).
Akenson, D.H. *God's Peoples: Covenant and Land in South Africa, Israel and Ulster* (Ithaca 1992).
Alexander, E. *Primate Alexander; Archbishop of Armagh* (London 1913).
Allen, R. *The Presbyterian College, Belfast: Centenary Volume 1853–1953* (Belfast 1954).
Armstrong, T. *My Life in Connaught, with Sketches of Mission Work in the West* (London 1906).
Bailey, A. *Acts of Union: Reports on Ireland 1973–79* (London 1980).
Barkley, J.M. *Blackmouth and Dissenter* (Belfast 1991).
Barnes, S. *All for Jesus: The Life of W.P. Nicholson* (Belfast 1996).
Barritt, D.P. and Carter, C.F. *The Northern Ireland Problem: A Study in Group Relations* (London 1962).
Barron, R. *The God of My Life: An Autobiography* (Belfast n.d. [1928?]).
Bartlett, T. and Jeffery, K. (eds) *A Military History of Ireland* (Cambridge 1996).
Bauman, Z. *Mortality, Immortality and other Life Strategies* (Cambridge 1992).
Bebbington, D.W. *Evangelicalism in Modern Britain: A History from the 1730s to the 1980s* (London 1995).
Beckett, I.F.W. (ed.) *The Army and the Curragh Incident 1914* (London 1986).
Belfrage, S. *Living with War: A Belfast Year* (New York 1987).
Bell, G.K.A. *Randall Davidson, Archbishop of Canterbury* (London 1952).
Bell, J.B. *Back to the Future: The Protestants and a United Ireland* (Dublin 1996).
Bell, M. *A History of Scouting in Northern Ireland* (Belfast 1985).
Bell, P.M.H. *Disestablishment in Ireland and Wales* (London 1969).
Bence-Jones, M. *Twilight of the Ascendancy* (London 1987).
Benson, A.C. *The Life of Edward White Benson* (London 1901).
Bernard, J.H. *In War Time* (London 1917).
Bew, P., Darwin, K. and Gillespie, G. (eds) *Passion and Prejudice: Nationalist–Unionist Conflict in Ulster in the 1930s and the Founding of the Irish Association* (Belfast 1993).
Biggs-Davidson, J. and Chowdharay-Best, G. *The Cross of St Patrick: The Catholic Unionist Tradition in Ireland* (Bourne End, Bucks 1984).
Birmingham, G.A. *Pleasant Places* (London 1934).
Birrell, A. *Things Past Redress* (London 1937).
Birrell, D. and Murie, A. *Policy and Government in Northern Ireland: Lessons of Devolution* (Dublin 1980).
Blake, R. *The Unknown Prime Minister: The Life and Times of Andrew Bonar Law 1858–1923* (London 1955).
Blanchard, P. *The Irish and Catholic Power* (London 1954).
Bowe, N.G. *The Life and Work of Harry Clarke* (Blackrock, 1994).
Bowen, D. *History and the Shaping of Irish Protestantism* (New York 1995).
Bowen, K. *Protestants in a Catholic State: Ireland's Privileged Minority* (Kingston, Ontario 1983).
Boyce, D.G. *Englishmen and Irish Troubles: British Public Opinion and the Making of Irish Policy 1918–22* (London 1972).
Boycott, C.A. *Boycott: The Life Behind the Word* (Ludlow 1997).

Boyd, R. *Manchuria and Our Mission There* (Belfast 1908).
Boyd, R.H. *The Path of Valour* (Belfast 1943).
Bradley, I. *Abide with Me: The World of Victorian Hymns* (London 1997).
Braley, E.F. (ed.) *Letters of Herbert Hensley Henson* (London 1951).
Brodie, M. *The Tele: A History of the Belfast Telegraph* (Belfast 1995).
Brown, A.J. *The Word of God among all Nations: The History of the Trinitarian Bible Society 1831–1981* (London 1981).
Buckland, P. *Irish Unionism 1885–1923: A Documentary History* (Belfast 1973).
Buckland, P. *A History of Northern Ireland* (Dublin 1981).
Bullock S.F. *After Sixty Years* (London n.d. [1931]).
Burns, D. *Temperance in the Victorian Age* (London 1897).
Burrows, E. and Mayes, P. *Specially Concerned – The Mothers' Union in Ireland 1887–1897* (Dublin 1986).
Butler, H. *The Sub-Prefect Should Have Held his Tongue, and Other Essays* (London 1990).
Butler, R.A. *The Art of the Possible* (London 1971).
Campbell, F. *The Dissenting Voice: Protestant Democracy in Ulster from Plantation to Partition* (Belfast 1991).
Carmichael, A. *Thou Givest ... They Gather* (London 1959).
Cathcart, R. *The Most Contrary Region: The BBC in Northern Ireland 1924–1984* (Belfast 1984).
Chadwick, O. *The Victorian Church*, Part I (London 1966); Part II (London 1970).
Chadwick, O. *Michael Ramsey: A Life* (Oxford 1991).
Clayton, P. *Enemies and Passing Friends: Settler Ideologies in Twentieth Century Ulster* (London 1996).
Clow, W.M. *Christ in the Social Order* (London 1913).
Cole, R.L. *History of Methodism in Ireland 1860–1960* (Belfast 1960).
Coles, M. *I Will Build My Church: The Story of the Congregational Union of Ireland 1829–1979* (Belfast 1979).
Colley, L. *Britons: Forging the Nation 1707–1837* (London 1992).
Collis, J. *The Sparrow hath Found Herself a House* (Dublin 1943).
Conroy, G. *Occasional Sermons, Addresses and Essays* (Dublin 1888).
Coogan, T.P. *De Valera: Long Fellow, Long Shadow* (London 1993).
Cooke, D. *Persecuting Zeal: A Portrait of Ian Paisley* (Dingle 1996).
Cooney, D.A.L. *A History of the Christian Endeavour Movement in Ireland* (Belfast 1977).
Corkey, E. *David Corkey: A Life Story of Work in the Slums, in a Country Parish and on the Battlefield* (London n.d. [1925]).
Corkey, W. *Glad Did I Live* (Belfast 1963).
Currie, R. *Methodism Divided* (London 1968).
D'Arcy, C.F. *The Adventures of a Bishop* (London 1934).
Davey, J.E. *A Memoir of the Revd Charles Davey* (Belfast 1921).
Davie, G. *Religion in Britain since 1945: Believing without Belonging* (Oxford 1994).
Davies, H. *Worship and Theology in England, Volume IV: From Newman to Martineau* (Grand Rapids 19976).
Dillon, M. *God and the Gun: The Church and Irish Terrorism* (London 1998).
Dudley Edwards, R. *The Faithful Tribe: An Intimate Portrait of the Loyal Institutions* (London 1999).

Dunlop, J. *A Precarious Belonging: Presbyterians and the Conflict in Ireland* (Belfast 1995).
Eames, R. *Chains to be Broken: A Personal Reflection on Northern Ireland and its People* (Belfast 1993).
Ellis, I. *Vision and Reality: A Survey of Twentieth Century Inter-Church Relations* (Belfast 1992).
Ellison, J. and Walpole, G.H.S. (eds) *Church and Empire: A Series of Essays on the Responsibilities of Empire* (London 1907).
Ervine, J. *Reminiscences* (Belfast 1911).
Ervine, St J. *Sir Edward Carson and the Ulster Movement* (London 1915).
Ervine, St J. *Craigavon: Ulsterman* (London 1949).
Farmar, T. *Ordinary Lives: Three Generations of Irish Middle Class Experience: 1907, 1932, 1963* (Dublin 1991).
Farren, S. *The Politics of Irish Education 1920–65* (Belfast 1995).
Faulkner, B. *Memoirs of a Statesman* (London 1978).
Feeney, J. *John Charles McQuaid: The Man and the Myth* (Dublin 1964).
Ferguson, N. *The Pity of War* (London 1998).
Figgis, T.F. and Drury, T.W.E. *Rathmines School Roll* (Dublin 1932).
Fisk, R. *In Time of War: Ireland, Ulster and the Price of Neutrality 1939–45* (London 1995).
FitzGerald, B. *Father Tom: An Authorised Portrait of Cardinal Tomas O Fiaich* (London 1990).
Fitzpatrick, D. *Oceans of Consolation: Personal Accounts of Irish Migration to Australia* (Cork 1994).
Fitzpatrick, D. *The Two Irelands 1912–1939* (Oxford 1998).
Fleming, L. *Head or Harp?* (London 1965).
Foster, R.F. *Modern Ireland 1600–1972* (London 1988).
Fulton, A. *Through Earthquake, Wind and Fire: Church and Mission in Manchuria 1867–1950* (Edinburgh 1967).
Fulton, A. *J. Ernest Davey* (Belfast 1970).
Gairdner, W.H.T. *'Edinburgh 1910': An Account and Interpretation of the World Missionary Conference* (Edinburgh 1910).
Gallagher, E. *At Points of Need: The Story of the Belfast Central Mission 1889–1989* (Belfast 1989).
Gallagher, E. and Worrall, S. *Christians in Ulster 1968–1980* (Oxford 1982).
Galliher, J.F. and DeGregory, J.L. *Violence in Northern Ireland: Understanding Protestant Perspectives* (Dublin 1985).
Galloway, P. *The Cathedrals of Ireland* (Belfast 1992).
Gibbs, M.E. *The Anglican Church in India 1600–1970* (Delhi 1972).
Gilbert, A.D. *The Making of Post-Christian Britain: A History of the Secularization of Modern Society* (London 1980).
Gillespie, R. and Kennedy, B.P. (eds) *Ireland: Art into History* (Dublin 1994).
Good, J.W. *Ulster and Ireland* (Dublin 1919).
Graham, B. *Just As I Am: The Autobiography of Billy Graham* (London 1997).
Gray, T. *The Orange Order* (London 1972).
Greacen, R. *The Sash my Father Wore: An Autobiography* (Edinburgh 1997).
Green, S.J.D. *Religion in the Age of Decline* (Cambridge 1996).
Grindle, W.H. *Irish Cathedral Music: A History of Music at the Cathedrals of the Church of Ireland* (Belfast 1989).

Grubb, G.W. *The Grubbs of Tipperary: Studies in Heredity and Character* (Cork 1972).
Grubb, I. *Quakers in Ireland 1654–1900* (London 1927).
Hamilton, T. *History of Presbyterianism in Ireland* (Belfast 1992).
Harbinson, R. *No Surrender: An Ulster Childhood* (London 1960).
Harford, J.B. and MacDonald, F.C. *Handley Carr Glyn Moule, Bishop of Durham: A Biography* (London [1922]).
Harford-Battersby, C.F. *Pilkington of Uganda* (London 1898).
Harkness, D. *Ireland in the Twentieth Century: Divided Island* (London 1996).
Harmon, M. *Sean O'Faolain* (London 1994).
Harte, F.E. *The Road I Have Travelled: The Experiences of an Irish Methodist Minister* (Belfast n.d. [1945]).
Hartford, D.M. *Among the Clergy* (Portlaw 1973).
Hartford, R.R. *Godfrey Day: Missionary, Pastor and Primate* (Dublin 1940).
Hastings, A. *A History of English Christianity 1920–1990* (London 1991).
Hatton, H.E. *The Largest Amount of Good: Quaker Relief in Ireland 1654–1921* (Kingston, Ontario 1993).
Hennessey, T. *A History of Northern Ireland 1920–1996* (London 1997).
Hime, M.C. *Home Education: or Irish Versus English Grammar Schools for Irish Boys* (London 1887).
Holmes, J. and Urquhart, D. *Coming into the Light: The Work, Politics and Religion of Women in Ulster 1840–1940* (Belfast 1994).
Holmes R.F.G. and Knox, R.B. (eds) *The General Assembly of the Presbyterian Church in Ireland 1840–1990* (Belfast 1990).
Houghton, F. *Amy Carmichael of Dohnavur* (London 1953).
Hughes, D.P. *The Life of Hugh Price Hughes* (London 1904).
Hughes, E. (ed.) *Culture and Politics in Northern Ireland 1960–1990* (Buckingham 1991).
Hurley, M. (ed.) *Irish Anglicanism* (Dublin 1970).
Hylson-Smith, K. *High Churchmanship in the Church of England* (Edinburgh 1993).
Inglis, B. *West Briton* (London 1962).
Iremonger, F.A. *William Temple, Archbishop of Canterbury: His Life and Letters* (London 1949).
Irvine, M. *Northern Ireland: Faith and Faction* (London 1991).
Jeffery, F. *Irish Methodism: An Historical Account of its Traditions, Theology and Influence* (Belfast 1964).
Jeffery, K. (ed.) *'An Irish Empire'? Aspects of Ireland and the British Empire* (Manchester 1996).
Johnstone, T.M. *The Vintage of Memory* (Belfast 1943).
Keane, F. *Letter to Daniel: Despatches from the Heart* (London 1996).
Kelly, W.R. *Firm and Deep: An Historical Account of the Formation of the 1st Belfast (1st Irish) Company, The Boys' Brigade* (Belfast 1978).
Kennedy, D. *The Widening Gulf: Northern Attitudes to the Independent Irish State 1919–49* (Belfast 1988).
Kennedy L. and Ollerenshaw P. *An Economic History of Ulster 1820–1939* (Manchester 1985).
Kenny, M. *Goodbye to Catholic Ireland* (London 1997).
Kerr, W.S. *Rev Andrew Boyd MA – A Memoir* (Dublin 1927).
King, S.H. and McMahon, S. *Hope and History: Eyewitness Accounts of Life in Twentieth-century Ulster* (Belfast 1995).

Kitson Clark, G. *An Expanding Society: Britain 1832–1900* (Cambridge 1967).
Kohn, L. *The Constitution of the Irish Free State* (London 1932).
Larkin, E. *James Larkin 1876–1947: Irish Labour Leader* (London 1965).
Larmour, P. *The Arts and Crafts Movement in Ireland* (Belfast 1992).
Lee, J.J. *Ireland 1912–1985: Politics and Society* (Cambridge 1989).
Legerton, H.J.W. *For our Lord and His Day: A History of the Lord's Day Observance Society* (Leicester [1979]).
Leslie, S. *Long Shadows* (London 1966).
Longmate, N. *The Water-Drinkers: A History of Temperance* (London 1968).
Macaulay, J. *Ireland in 1872* (London 1873).
McCarthy, M.J.F. *Priests and People in Ireland* (Dublin 1902).
McCarthy, M.J.F. *Five Years in Ireland 1895–1900* (Dublin 1903).
McCartney, D.J. *Nor Principalities Nor Powers: A History (1621–1991) of 1st Presbyterian Church, Carrickfergus* (Belfast 1991).
McCaughey, T.P. *Memory and Redemption: Church, Politics and Prophetic Theology in Ireland* (Dublin 1993).
McCrea, A. (ed.) *Irish Methodism in the Twentieth Century; A Symposium* (Belfast 1931).
McCreedy, J. *The Seer's House: The Remarkable Story of James McConnell and the Whitewell Metropolitan Tabernacle* (Belfast 1997).
McCrum, M. *The Craic: A Journey Through Ireland* (London 1998).
McDermott, R.P. and Webb, D.A. *Irish Protestantism Today and Tomorrow* (Dublin n.d. [1937]).
McDowell, R.B. *The Church of Ireland 1869–1969* (London 1975).
McDowell, R.B. and Webb, D.A. *Trinity College, Dublin 1592–1952: An Academic History* (Cambridge 1982).
McEwan, J.M. (ed.) *The Riddell Diaries 1908–1923* (London 1986).
McIvor, B. *Hope Deferred: Experiences of an Irish Unionist* (Belfast 1998).
Mackail, J.A. and Wyndham, G. *Life and Letters of George Wyndham* (2 vols, London n.d. [1924]).
McKinney, J.W. and Sterling, W.S. *Gurteen College: A Venture of Faith* (Omagh 1972).
McMinn, J.R.B. *Against the Tide: A Calendar of the Papers of Rev J.B. Armour* (Belfast 1985).
MacNeice, J.F. *Some Northern Churchmen and Some Notes on the Church in Belfast* (Belfast 1934).
MacStiofáin, S. *Revolutionary in Ireland* (Farnborough 1974).
Magee, H. *A Short Account of the History and Methods of the Assembly's Mission in Dublin* (Dublin 1885).
Maguire, E. *Fifty-Eight Years of Clerical Life in Ireland: An Autobiography* (Dublin 1904).
Mahto, S. *Hundred Years of Christian Missions in Chotanagpur since 1845* (Ranchi 1971).
Malcolm, E. *'Ireland Sober, Ireland Free': Drink and Temperance in Nineteenth-Century Ireland* (Syracuse 1986).
Manning, B.L. *The Hymns of Wesley and Watts* (London 1960).
Mansergh, D. (ed.) *Nationalism and Independence: Selected Irish Papers by Nicholas Mansergh* (Cork 1997).
Maxwell, V. *Belfast's Halls of Faith and Fame* (Belfast 1999).

Mearns, A. *The Bitter Cry of Outcast London* (London 1883).
Megahey, A.J. *Humphrey Gibbs, Beleaguered Governor: Southern Rhodesia 1929–69* (London 1998).
Miller, K.A. *Emigrants and Exiles: Ireland and the Irish Exodus to North America* (New York 1988).
Milne, K. *The Church of Ireland: A History* (Dublin 1966).
Mission Completed: T.S. Mooney of Londonderry 1907–86 (Lisburn [1986]).
Moloney, E. and Pollak, A. *Paisley* (Swords 1986).
Monypenny, W.F. *The Two Irish Nations: An Essay on Home Rule* (London 1913).
Moody, A.F. *Memories and Musings of a Moderator* (London n.d. [1938]).
Moody, T.W. and Beckett, J.C. *Queen's, Belfast 1845–1949: The History of a University* (2 vols, London 1959).
Moore, H.K. *Reminiscences and Reflections from some Sixty Years of Life in Ireland* (London 1930).
Morgan, A. *Labour and Partition: The Belfast Working Class 1905–23* (London 1991).
Munson, J. *The Nonconformists: In Search of a Lost Culture* (London 1991).
Murphy, D. *A Place Apart* (London 1978).
Murray, R.H. *Archbishop Bernard: Professor, Prelate and Provost* (London 1931).
Newman, J. *Change and the Catholic Church* (Baltimore 1965).
Nicholas, W. *Christianity and Socialism* [23rd Fernley Lecture] (London 1893).
Nicolson, H. *Helen's Tower* (London 1937).
Noll, M.A. *A History of Christianity in the United States and Canada* (London 1992).
Norman, E.R. *The Catholic Church in Ireland in the Age of Rebellion 1859–1873* (London 1965).
Norman, E.R. *Church and Society in England 1770–1970* (Oxford 1976).
Norman, E.R. *Christianity and the World Order: The BBC Reith Lectures 1978* (Oxford 1979).
O'Connell, E.E. *Northern Irish Stereotypes* (Dublin 1977).
O'Connor, F. *In Search of a State: Catholics in Northern Ireland* (Belfast 1993).
O'Connor, T.C. *The Image of a Cross in Pagan, Christian and Anti-Christian Symbolism* (Dublin 1894).
O'Dowd, M. and Wichert, S. (eds) *Chattel, Servant or Citizen: Women's Status in Church, State and Society* (Belfast 1995).
O'Driscoll, H. *The Leap of the Deer: Memories of a Celtic Childhood* (Boston, Mass. 1994).
O'Neill, Lord *The Autobiography of Terence O'Neill* (London 1972).
Orr, P. *The Road to the Somme: Men of the Ulster Division Tell Their Story* (Belfast 1987).
Parker, O. *For the Family's Sake: A History of the Mothers' Union 1876–1976* (London 1975).
Patton, H.E. *Fifty Years of Disestablishment: A Sketch* (Dublin 1922).
Paul, L. *The Deployment and Payment of the Clergy* (London 1964).
Peabody, F.G. *Jesus Christ and the Social Question* (New York 1901).
Quinn, C. *et al. Directions: Theology in a Changing Church* (Dublin 1970).
Rattenbury, J.E. *The Eucharistic Hymns of John and Charles Wesley* (London 1948).
Redmond, J. *Church, State, Industry 1827–1929 in East Belfast* (Belfast 1961).
Richardson, N. (ed.) *A Tapestry of Beliefs: Christian Traditions in Northern Ireland* (Belfast 1998).
Rigg, J.H. *A Comparative View of Church Organisation* (London 1897).

Rolleston, C.H. *Portrait of an Irishman: A Biographical Sketch of T.W.O. Rolleston* (London 1939).
Rutherford, J.S. *Christian Reunion in Ireland* (Dublin [1942]).
Seaver, G. *John Allen Fitzgerald Gregg, Archbishop* (London 1963).
Seddall, H. *The Church of Ireland: A Historical Sketch* (Dublin 1886).
Selborne, Earl of *A Defence of the Church of England against Disestablishment* (London 1886).
Sibbett, R.M. *For Christ and Crown: The Story of a Mission* (Belfast 1926).
Smiley, M.H.H. *The Life and Letters of the Revd W. Smiley* (London 1888).
Smith, C.F. *James Nicholson Richardson of Bessbrook* (London 1925).
Springhall, J. (ed.) *Sure and Stedfast: A History of the Boys' Brigade 1883–1983* (London 1983).
Springhall, J. *Coming of Age: Adolescence in Britain 1860–1960* (Dublin 1986).
Stanford, W.B. and McDowell, R.B. *Mahaffy: A Biography of an Anglo-Irishman* (London 1975).
Stewart, A.T.Q. *The Ulster Crisis* (London 1967).
Stewart, A.T.Q. *The Narrow Ground: Aspects of Ulster 1609–1969* (London 1977).
Stringer, P. and Robinson, G. *Social Attitudes in Northern Ireland* (Belfast 1991).
Sykes, N. *Man as Churchman [The Wiles Lectures 1959]* (Cambridge 1960).
Taggart, N.W. *The Irish and World Methodism 1760–1900* (London 1986) 1980).
Thompson, J. (ed.) *Into all the World: A History of 150 years of the Overseas work of the Presbyterian Church in Ireland* (Belfast 1990).
Tolhurst, J. *A Companion to 'Veritas Splendor'* (Leominster 1994).
Troeltsch, E. *The Social Teaching of the Christian Churches* (2 vols, London 1956).
Utley, T.E. *Lessons of Ulster* (London 1973).
Waddell, H.C. *John Waddell* (Belfast 1949).
Warke, R.A. *Ripples in the Pool* (Dublin 1992).
White, T. de V. *The Anglo-Irish* (London 1972).
Whitelaw, W. *The Whitelaw Memoirs* (London 1989).
Whiteside, L. *George Otto Simms, A Biography* (Gerards Cross 1990).
Wilkinson, A. *The Church of England and the First World War* (London 1978).
Wilson, B.R. (ed.) *Patterns of Sectarianism* (London 1967).
Winchester, S. *In Holy Terror: Reporting the Ulster Troubles* (London 1974).

Reports, articles, pamphlets, theses

Only those cited in the Notes are listed here.

Administration 1967 (Dublin 1967).
Agrawal, A. *et al. A Christian Response to the Irish Situation* (Belfast 1997).
Anderson, J. and Shuttleworth, I. 'Sectarian Readings of Sectarianism: Interpreting the Northern Ireland Census', *Irish Review*, no 16 (Autumn/Winter 1994), pp. 74–93.
Baker, S. 'Orange and Green: Belfast 1832–1912', *The Victorian City: Images and Realities Volume 2*, ed. H.J. Dyos and M. Wolff (London 1973), pp. 789–818.
Barkley, J.M. 'Marriage and the Presbyterian Tradition', *Ulster Folklife*, vol 39 (1993), pp. 29–40.

Bebbington, D.W. 'The City, the Countryside and the Social Gospel in Late Victorian Nonconformity', in *The Church in Town and Countryside*, ed. D. Baker (Oxford 1979), pp. 415–26.

Berger, P.L. 'Against the Current', *Prospect*, March 1997.

Bernard, J.H. *The Present Position of the Irish Church* (London 1904).

Blair, S.A. 'The Rev W.F. Marshall: Profile of an Ulsterman', *Bulletin of the Presbyterian Historical Society of Ireland*, no 24 (1995), pp. 1–20.

Bleakley, D. *Young Ulster and Religion in the Sixties* (Belfast 1964).

Bradshaw, J. *Blood Worship: Christian Idolatory: the fallacies and heresies of our hymnology* (Belfast n.d. [1914?]).

Brown, K.D. 'Life after Death? A Preliminary Survey of the Irish Presbyterian Ministry in the Nineteenth Century', *Irish Economic and Social History*, xxii (1995), pp. 49–63.

Brown, L.T. 'The Presbyterian Dilemma: A Survey of the Presbyterians and Politics in Counties Cavan and Monaghan over Three Hundred Years', *Clogher Record*, vol XV, no 2 (1995), pp. 30–68.

Brown, R. 'Time to Rethink?', *Aspects of Education*, no 52 (1995), pp. 50–8.

Buchanan, A. 'In Retrospect: Alan Alexander Buchanan', *Search*, vol 17, no 1 (Spring 1994), pp. 34–9.

Butler, J. 'Select Documents XLV: Lord Oranmore's Journal 1913–27', *Irish Historical Studies*, vol xxix, no 116 (November 1995), pp. 553–93.

Canavan, J.E. 'The Future of Protestantism in Ireland', *Studies*, vol xxxiv, no 134 (June 1945), pp. 23–40.

Carson, H.M. *Riots and Religion* (Worthing n.d. [1969]).

Collins, J. 'Political Attitudes and Integrated Education in the North of Ireland', *Studies*, vol 79, no 316 (Winter 1990), pp. 409–19.

Compton, P.A. 'An Evaluation of the Changing Religious Composition of the Population of Northern Ireland', *Economic and Social Review*, vol 16, no 3 (April 1985), pp. 201–24.

Compton, P.A. 'Demography – The 1980s in Perspective', *Studies*, vol 80, no 318 (Summer 1991), pp. 157–68.

Considère-Charon, M-C. 'The Church of Ireland: Continuity and Change', Studies, vol 87, no 346, pp. 107–16.

Cooke, J.H. 'Church Methodists in Ireland: Statistical Evidence', *Proceedings of the Wesley Historical Society*, vol xxxiv, Part 6 (June 1964), pp. 135–40.

Corkey, W. *The McCann Mixed Marriage Case* (Edinburgh 1911).

Corkey, W. *Episode in the History of Protestant Ulster 1923–1947* (Belfast [1959]).

Daly, C.B. 'A Vision of Ecumenism in Ireland', *Eire-Ireland*, vol 17, part 1 (Spring 1982), pp. 7–30.

Davey, J.E. *1840–1940: The Story of a Hundred Years ... of the Irish Presbyterian Church* (Belfast 1940).

Davis, R. 'Nelson Mandela's Irish Problem: Republican and Loyalist Links with South Africa 1970–1990', *Eire-Ireland*, 27 (1992), pp. 47–68.

Devlin, C.A. 'The Eucharistic Procession of 1908: The Dilemma of the Liberal Government', *Church History*, vol 63, no 3 (September 1994), pp. 407–25.

Dooley, T.A.M. 'Monaghan Protestants in a Time of Crisis, 1919–22', in *Religion, Conflict and Coexistence in Ireland*, ed. R.V. Comerford *et al*, (Dublin 1990), pp. 235–43.

Dooley, T.A.M. 'The Organisation of Unionist Opposition to Home Rule in Counties Monaghan, Cavan and Donegal, 1885–1912', *Clogher Record*, vol xvi, no 1 (1997), pp. 46—70.

Douglas, I.T. '"A Light to the Nations": Episcopal Foreign Missions in Historical Perspective', *Anglican and Episcopal History*, vol lxi, no 4 (December 1992), pp. 449–81.

Dowling, M. 'A History of the Irish Baptist College', *Irish Baptist Historical Society Journal*, vol 13 (1980–81), pp. 29–41.

Elliott, M. 'Religion and Identity in Northern Ireland', in W.A. Van Horne, *Global Convulsions: Race, Ethnicity and Nationalism at the End of the Twentieth Century* (Albany 1997), pp. 149–67.

Facts showing the Progress and Present Position of Ritualism in Ireland (Dublin 1890).

Faughnan, S. 'The Jesuits and the Drafting of the Irish Constitution of 1937', *Irish Historical Studies*, vol xxvi, no 101 (May 1988), pp. 79–102.

First Report of the Royal Commission on University Education Ireland (Robertson Commission, 1902) [Cd. 825–6].

Fitzpatrick, D. 'The Logic of Collective Sacrifice: Ireland and the British Army 1914–18', *Historical Journal*, 38, 4 (December 1995) pp. 1017–30.

Flanagan, K. 'The Shaping of Irish Anglican Secondary Schools, 1854–1878', *History of Education*, vol 13, no 1 (1984), pp. 27–43.

Free Church Year Book and Official Report of the Fifth National Council of the Evangelical Free Churches ... 1900 (London 1900).

French, R.B.D. 'J.O. Hannay and the Gaelic League', *Hermathena*, no cii (Spring 1966), pp. 26–52.

Gallagher, R.D.E. 'Northern Ireland – The Record of the Churches', *Studies*, vol 80, no 318 (Summer 1991), pp. 169–77.

Gilley, S. 'Religion in Modern Ireland', *Journal of Ecclesiastical History*, vol 43, no 1 (January 1992), pp. 110–15.

Gregg, J.A.F. *Reunion: A Paper read to the Clergy of the Diocese of Glasgow* (Published by the Council for the Defence of Church Principles, 1949).

Hammond, T.C. 'The Religious Question in Ireland', *Nineteenth Century*, vol lxxiii, no 432 (February 1913), pp. 338–54.

Hartin, J. 'The Challenge to Irish Ecumenism', *Studies*, vol lxxii, no 288 (Winter 1983), pp. 291–308.

Heritage and Renewal: The Report of the Archbishops' Commission on Cathedrals (London 1994).

Hilliard, D. 'Unenglish and Unmanly: Anglo-Catholicism and Homosexuality', *Victorian Studies*, vol 25, no 2 (Winter 1982), pp. 170–210.

Historical Sketch of the First Church of Christ, Scientist, Belfast (Belfast 1944).

Holden, D.E.W. 'Will the Inward Light go out in Ireland?', *Journal of the Society of Friends*, vol 56, no 4, pp. 311–27.

Holmes, R.F.G. *Presbyterians and Orangeism 1795–1995* (Belfast 1996).

Jackson, A. 'Irish unionism and the Russellite threat, 1894–1906', *Irish Historical Studies*, vol xxv, no 100 (November 1987), pp. 376–404.

Jackson, A. 'Unionist Politics and Protestant Society in Edwardian Ireland', *Historical Journal*, 33, 4 (1990), pp. 839–66.

Jarman, N. 'Displaying Faith: The Development of Ulster's Banner Parading Tradition', *New Ulster*, no 23 (Summer 1994), pp. 10–12.

Jeffery, F. 'Church Methodists in Ireland', *Proceedings of the Wesley Historical Society*, vol xxxiv, Part 4 (December 1963), pp. 73–5.

Jeffery, F. 'Four Methodist Hymn Books', *Bulletin of the Wesley Historical Society (Irish Branch)*, vol 3, part 1 (Autumn 1996), pp. 14–19.

Jones, V. 'The Attitudes of the Church of Ireland Board of Education to Textbooks in National Schools 1922–67', *Irish Educational Studies*, vol 11, no 1 (1992), pp. 72–81.

King, R.G.S. *Ulster's Refusal to Submit to a Roman Catholic Parliament Stated and Justified* (Derry 1914).

Knox, R.B. 'The Irish Contribution to English Presbyterianism', *Journal of the United Reformed Church Historical Society*, vol 4, no 1 (October 1987), pp. 22–35.

Lambeth Conference 1930: Encyclical Letter from the Bishops with the Resolutions and Reports (London 1930).

Lambeth Conference 1948: Encyclical Letter from the Bishops with the Resolutions and Reports (London 1948).

Lapsley, D. 'Where is the Ecumenical Movement Going Today?', *Studies*, vol lxxii, no 288 (Winter 1983), pp. 299–308.

Larkin, E. 'Church, State and Nation in Modern Ireland', *American Historical Review*, vol 80, no 5 (December 1975), pp. 1244–76.

Lee, R.M. 'Intermarriage, Conflict and Social Control in Ireland: The Decree "*Ne temere*"', *Economic and Social Review*, vol 17, no 1 (October 1985), pp. 11–27.

Livingstone, P. *The Fermanagh Story* (Enniskillen 1969).

Macaulay, J. *The Price of our Presbyterianism: A Historical Sketch of the Sustentation Fund of the Presbyterian Church in Ireland* (Belfast 1939).

McCann, J. 'Christian Education in the Service of Reconciliation', *Aspects of Education*, no 52 (1995), pp. 66–74.

McClelland, A. 'Orange Folk Art in Ireland', *Ulster Folklife*, vol 42 (1996), pp. 12–22.

McDermott, R.P. 'Church and State in Northern Ireland', *Theology*, vol lxxii, no 586 (April 1969), pp159–69.

Machin, G.I.T. 'The Last Victorian Anti-Ritualist Campaign, 1895–1906', *Victorian Studies*, vol 25, no 3 (Spring 1982), pp. 277–302.

McKee, E. 'Church–State Relations and the Development of Irish Health Policy: The Mother-and-Child scheme, 1944–53', *Irish Historical Studies*, vol xxv, no 98 (November 1986), pp. 159–94.

MacVeagh, J. *Home Rule or Rome Rule* (London 1912).

McVeigh, R. 'The Specificity of Irish Racism', *Race and Class*, vol 33, no 4 (April–June 1992), pp. 31–45.

Maguire, M. 'A Socio-Economic Analysis of the Dublin Protestant Working Class, 1870–1926', *Irish Economic and Social History*, vol xx (1993), pp. 35–61.

Mahaffy, J.P. 'Provincial Patriotism', *Nineteenth Century*, vol xxxvii, no ccxx (June 1895), pp. 1027–35.

Mahaffy, J.P. 'Will Home Rule be Rome Rule?', *Blackwood's Magazine*, vol cxcii, no mclxii (August 1912), pp. 153–9.

Marriage and the Family in Ireland To-day: A Social and Pastoral Study (Belfast 1987).

Mathias, P. 'The Brewing Industry, Temperance and Politics', *Historical Journal*, vol 1, no 2 (1958), pp. 97–114.

Miller, D.W. 'Presbyterianism and "Modernization" in Ulster', *Past and Present*, no 80 (August 1978), pp. 66–90.

Milne, K. *Protestant Aid 1836–1986: A History of the Association for the Relief of Distressed Protestants* (Dublin 1989).
Morgan, P.P. 'A Study of the Work of American Revivalists in Britain from 1870–1914 and the Effect upon Organized Christianity of their Work there' (BLitt thesis, Oxford, 1961).
Morgan, V. and Fraser, G. 'Integrated Schools, Religious Education and the Churches in Northern Ireland', *Aspects of Education*, no 52 (1995), pp. 30–9.
Morris, G. O'C. 'The Dublin University Mission to Chota Nagpur: A Sketch of its Work', *Irish Church Quarterly*, vol vi, no 24 (October 1913), pp. 302–13.
Murray, J.W. *How can we make our Churches and Services more attractive to the people* (Dublin 1868).
Newsinger, J. '"I bring not peace but a sword": The Religious Motif in the Irish War of Independence', *Journal of Contemporary History*, vol 13, no 3 (July 1978), pp. 609–28.
Norris, D. 'Homosexual People and the Christian Churches in Ireland: A Minority and its Oppressors', *The Crane Bag*, vol 5, no 1 (1981), pp. 31–7.
O'Farrell, P. 'Millenialism, Messianism and Utopianism in Irish History', *Anglo-Irish Studies*, ii (1976), pp. 45–68.
Paisley, I.R.K. *Are We to Lose our Protestant Heritage For ever?* (Belfast n.d.).
Paton, D.M. (ed.) *Breaking Barriers: Nairobi 1975* (London 1976).
Patterson, C. *Over the Hill: Ecumenism in the Irish Presbyterian Church* (Belfast 1997).
Patterson, H. 'Independent Orangeism and Class Conflict in Edwardian Belfast: A Reinterpretation', *Proceedings of the Royal Irish Academy*, vol 80c (1980), pp. 1–27.
Pooler, L.A. 'The Socialist Movement', *Irish Church Quarterly*, vol v, no 20 (October 1912), pp. 294–303.
Pyper, J. *The Irish Sacramental Wine Association* (Belfast 1875).
Stanford, W.B. *A Recognised Church: The Church of Ireland in Eire* (Dublin 1944).
Stevens, D. 'Dealing with the Lethal Toxin of Sectarianism', *Search*, vol 17, no 1 (Spring 1994), pp. 7–16.
The Growth and Development of the Girls' Brigade Ireland 1893–1983 (Belfast n.d. [1983?]).
'The Irish University Question', *Quarterly Review*, vol clxxxvii, no 374 (April 1898), pp. 567–89.
The Lambeth Conference 1958: The Encyclical Letter from the Bishops together with the Resolutions and Reports (London 1958).
The Report of the Lambeth Conference 1978 (London 1978).
Thompson, J. 'The Origin of the Irish Baptist Foreign Mission', *Irish Baptist Historical Society Journal*, vol 1 (1968/69), pp. 17–35.
Thompson, J. 'Irish Baptist College List of Students from 1892–1963', *Irish Baptist Historical Society Journal*, vol 23 (1990–91), pp. 39–44.
Thompson, J. 'Irish Baptist College List of Students from 1964–1991', *Irish Baptist Historical Society Journal*, vol 24 (1991–92), pp. 61–9.
Thompson, J. 'The Churches in Northern Ireland – Problem or Solution? A Presbyterian Perspective', *Irish Theological Quarterly*, vol 58, no 4 (1992), pp. 264–75.
Towards Reconciliation: The Interim Statement of the Anglican–Methodist Unity Commission (London 1967).
Violence in Ireland: A Report to the Churches (Belfast, revised edition, 1977).

Walker, G. '"Protestantism before Party!": The Ulster Protestant League in the 1930s', *Historical Journal*, 28, 4 (1985), pp. 961–7.

Walker, G. 'The Irish Presbyterian Anti-Home Rule Convention of 1912', *Studies*, vol 86, no 341, pp. 71–7.

Walsh, O. 'Protestant Female Philanthropy in Dublin in the Early 20th Century', *History Ireland*, (Summer 1997), pp. 27–31.

Warke, J. 'Baptists in Belfast: The Twentieth Century Challenge of Urban Growth and Decline', *Irish Baptist Historical Society Journal*, vol 20 (1987/88), pp. 12–19.

Wheatcroft, G. 'Ireland and the Left', *Prospect*, June 1998, pp. 10–11.

World Council of Churches *World Conference on Church and Society: Official Report* (Geneva 1967).

Index